ChatGPT for Accelerating Salesforce Development

Achieve faster, smarter, and more cost-effective Salesforce Delivery with ChatGPT

Andy Forbes

Philip Safir

Joseph Kubon

Francisco Fálder

BIRMINGHAM—MUMBAI

ChatGPT for Accelerating Salesforce Development

Group Product Manager: Alok Dhuri
Publishing Product Manager: Kushal Dave
Book Project Manager: Deeksha Thakkar
Senior Editor: Kinnari Chohan
Technical Editor: Vidhisha Patidar
Copy Editor: Safis Editing
Indexer: Hemangini Bari
Production Designer: Jyoti Kadam
DevRel Marketing Coordinator: Deepak Kumar and Mayank Singh

First published: December 2023

Production reference: 1211223

Published by Packt Publishing Ltd.
Grosvenor House
11 St Paul's Square
Birmingham
B3 1RB, UK

ISBN 978-1-83508-407-6

www.packtpub.com

In the pages that follow, my gratitude must first be extended to my family, who graciously shared me with this book. To Claudia, my partner in everything, your patience and support have been my anchor. To Grace and Chance, your laughter and love have been the light on this path. Your collective sacrifice allowed me to pursue the learning that has informed my writing and made this book possible. This book owes much to your tolerance of my divided attention and your encouragement. Thank you.

- Andy Forbes

To my loving wife and daughters, for sharing your joy, determination, and patience as we navigate life's adventures together. To my parents, who have instilled the perspective and confidence to take risks and challenge the status quo.

- Philip Safir

As I embarked on this endeavor, I took a moment to reflect on my Trailblazer Journey. The foundation of my passion was cemented by John Frum and Moises Cuellar. The insight and guidance you two infused in me is carried through this effort. #PelicanDown To my family, my children, Paige Ellen and Alex, and my wife Lisa, you three are the purpose of my life. The unconditional love, encouragement, and patience fuels my courage to be my authentic self. Thank you for providing me space to thrive. With you, there is nothing that cannot be conquered. I look forward to sharing all the wonders of life together. #LivingProof

- Joseph Kubon

A Santiago y Ana por darme la confianza y fuerzas para realizar nuevos proyectos. A mis padres y hermanas, por haberme dado forma, no sería el mismo sin vosotros. Os quiero.

- Francisco Fálder

We would like to thank the team at Packt for their sponsorship, guidance, and expertise throughout the creation of this book. Thank you to our technical reviewers whose expertise and feedback helped us iterate towards a quality final product. To the team at OpenAI, thank you for releasing a technology that demonstrates the promise and potential of generative AI. To our readers, we thank you for sharing a common interest in this topic and hope that this book serves you well in your Salesforce journey.

- Andy, Joseph, Paco, and Philip

Contributors

About the authors

Andy Forbes has an extensive career in technology, spanning over four decades, and is now channeling his IT experience towards investigating the impact of artificial intelligence, especially Generative AI, on Salesforce project delivery. His expertise in CRM and project management is complemented by his ITIL and Salesforce certifications. Having spent a decade at a Global Systems Integrator and driven by his entrepreneurial mindset, Andy has successfully led numerous Salesforce projects for Fortune 500 clients. His commitment to innovation and excellence is evident in his approach to enhancing IT service delivery and usage with the help of AI.

Philip Safir is a consulting executive and business architect. He serves enterprise clients via the delivery of technology roadmaps, process improvement, and solutions on the Salesforce platform. In his latest role as the Head of Salesforce Professional Services Delivery & Talent for a Global Systems Integrator, he was responsible for a team of 250 consultants and a $100M+ portfolio. His career spans Fortune 500, start-up, and international companies across various industry domains including Manufacturing, Retail, Financial Services, Telecom, and Non-Profit.

Joseph Kubon, an experienced Solution Architect for global enterprise deliveries, Salesforce MVP with 40+ Salesforce certifications and inventor (holding several patents), navigates the realms of manufacturing, health and media industries with a results-driven approach. Skilled in Agile methodologies, Business Process, and Architecture values, he carries a toolkit replete with Salesforce configuration and customization expertise for groundbreaking development. Joseph's analytical skills shine as he navigates software documentation, business intelligence, and design best practices. Guided by the wisdom that 'just because it can be built doesn't mean it should be', Joseph embraces the multiplicity of solutions to tomorrow's challenges measuring success with his "Time to Value" principles.

Francisco Fálder is a seasoned Salesforce maven and master of digital transformation. His career stands as a testament to a commitment to delivering top-tier, complex projects, seamlessly merging business and tech to deliver standout customer experiences. With every project, Paco re-imagines and redefines the digital landscape, fostering an environment where innovation is not only encouraged but celebrated. Passionate about the ever-evolving tech world, he has honed a deep-rooted affinity for Artificial Intelligence, Continuous Integration/Continuous Deployment (CI/CD), and Agile methodologies. These tools form the backbone of Paco's work, propelling them to create, innovate, and elevate in a rapidly shifting digital sphere. He is dedicated to unleashing the power of technology, one transformation at a time.

Table of Contents

4

Using ChatGPT for Salesforce Feature Design   65

5

Using ChatGPT for Salesforce Apex That Someone Else Wrote   85

6

Using ChatGPT for Salesforce Apex   103

7

Using ChatGPT for Salesforce Web Services and Callouts   125

8

Using ChatGPT for Salesforce Triggers   147

9

Using ChatGPT for Lightning Web Components   169

14

What You've Learned and What's Next   273

Appendix A: Case Study   279

Appendix B: A Deep Dive Into ChatGPT and User Stories   309

Index   327

Other Books You May Enjoy   332

Preface

The recent intersection of artificial intelligence with software applications and activities has opened new avenues for innovation and efficiency. This book is a testament to this exciting era, focusing on the integration of ChatGPT with Salesforce development—a synergy that is reshaping how we approach the delivery of customer relationship management features in profound ways.

This work is a collaborative effort among four co-authors, each contributing their distinct expertise and writing style. Due to our varied backgrounds, you may notice differences in language, formatting, and the level of detail in each chapter. These variations reflect our individual approaches and are part of the collaborative nature of this project. We have aimed to integrate these diverse perspectives into a coherent and informative guide. While each chapter may differ slightly in style, our collective goal remains to provide you with a useful and comprehensive resource. We appreciate your understanding as you encounter these variations and hope that the book serves as a valuable tool in your studies and professional endeavors.

Thank you for choosing our book, and we hope you find it informative and helpful.

Who this book is for

This book is for Salesforce business analysts, architects, developers, testers, and product owners. For each of these roles, the use of ChatGPT for Salesforce development offers a transformative approach to navigating and leveraging Salesforce's enormous capabilities and vast ecosystem.

What this book covers

Chapter 1, Getting Started with ChatGPT for Salesforce Development

Chapter 2, Using ChatGPT for Salesforce Configuration

Chapter 3, Using ChatGPT for Salesforce Flows

Chapter 4, Using ChatGPT for Salesforce Feature Design

Chapter 5, Using ChatGPT for Salesforce Apex That Someone Else Wrote

Chapter 6, Using ChatGPT for Salesforce Apex

Chapter 7, Using ChatGPT for Salesforce Web Services and Callouts

Chapter 8, Using ChatGPT for Salesforce Triggers

Chapter 9, Using ChatGPT for Salesforce Lightning Web Components

Chapter 10, Using ChatGPT for Salesforce Project Documentation

Chapter 11, Using ChatGPT for Salesforce User Stories

Chapter 12, Using ChatGPT for Salesforce Testing Scripts

Chapter 13, Using ChatGPT for Salesforce Debugging

Chapter 14, What You've Learned and What's Next

Appendix A, Case Study

Appendix B, A Deep Dive Into ChatGPT and User Stories

To get the most out of this book

While the the authors made an effort to make this book usable to the broadest possible audience, it does assume a basic understanding of Salesforce, the use epics and user stories, and how software projects are run. If you are at the very beginning of your Salesforce journey, we would recommend that you spend some time on the Salesforce Trailhead site: `https://trailhead.salesforce.com/`

Software/hardware covered in the book	Operating system requirements
Web Browser	Windows, MacOS, Linux
Salesforce	
ChatGPT	

- **For Salesforce, the latest stable version of Chrome is recommended.**

- **ChatGPT will work on Chrome as well.**

- **If you do not have access to a Salesforce org, you can signup for a free one here:** `https://developer.salesforce.com/signup`

- **If you do not have access to ChatGPT, you can signup for a free account here:** `https://chat.openai.com/auth/login`

Disclaimer

This book has been created by authors, technical experts, and a professional publishing team. We use many tools, including cutting-edge AI such as ChatGPT, to create the best possible material for our readers to help them on their IT journey.

Conventions used

There are a number of text conventions used throughout this book.

`Code in text`: Indicates code words in text, database table names, folder names, filenames, file extensions, pathnames, dummy URLs, user input, and Twitter handles. Here is an example: "Please write positive and negative test classes for the rewritten `AnalyzeContactDistance` class."

A block of code is set as follows:

```
@isTest
private class AnalyzeContactDistanceTestSetup {
    @TestSetup
    static void setupTestData() {
        Account acc = new Account(Name = 'Test Account',
BillingLatitude = 37.7749, BillingLongitude = -122.4194);
        insert acc;
```

Bold: Indicates a new term, an important word, or words that you see onscreen. For instance, words in menus or dialog boxes appear in **bold**. Here is an example: "Select **System info** from the **Administration** panel."

> **Tips or important notes**
> Appear like this.

Get in touch

Feedback from our readers is always welcome.

General feedback: If you have questions about any aspect of this book, email us at `customercare@packtpub.com` and mention the book title in the subject of your message.

Errata: Although we have taken every care to ensure the accuracy of our content, mistakes do happen. If you have found a mistake in this book, we would be grateful if you would report this to us. Please visit `www.packtpub.com/support/errata` and fill in the form.

Piracy: If you come across any illegal copies of our works in any form on the internet, we would be grateful if you would provide us with the location address or website name. Please contact us at `copyright@packtpub.com` with a link to the material.

If you are interested in becoming an author: If there is a topic that you have expertise in and you are interested in either writing or contributing to a book, please visit `authors.packtpub.com`.

Share Your Thoughts

Once you've read *ChatGPT for Accelerating Salesforce Development*, we'd love to hear your thoughts! Scan the QR code below to go straight to the Amazon review page for this book and share your feedback.

https://packt.link/r/1835084079

Your review is important to us and the tech community and will help us make sure we're delivering excellent quality content.

Download a free PDF copy of this book

Thanks for purchasing this book!

Do you like to read on the go but are unable to carry your print books everywhere?

Is your eBook purchase not compatible with the device of your choice?

Don't worry, now with every Packt book you get a DRM-free PDF version of that book at no cost.

Read anywhere, any place, on any device. Search, copy, and paste code from your favorite technical books directly into your application.

The perks don't stop there, you can get exclusive access to discounts, newsletters, and great free content in your inbox daily

Follow these simple steps to get the benefits:

1. Scan the QR code or visit the link below

https://packt.link/free-ebook/9781835084076

2. Submit your proof of purchase
3. That's it! We'll send your free PDF and other benefits to your email directly

Getting Started with ChatGPT for Salesforce Development

Welcome to a comprehensive and uniquely tailored reference guide designed to navigate the fascinating intersection of Salesforce projects and generative **artificial intelligence** (**AI**), powered by ChatGPT. This book is more than a linear journey; it's a flexible roadmap, created to serve both the novice and the seasoned professional who wants to harness the transformative capabilities of ChatGPT within the Salesforce ecosystem.

Unlike traditional guidebooks, you won't find it necessary to read this from cover to cover. Instead, it's structured as a modular guide, where each chapter stands on its own, focusing on specific aspects of Salesforce, such as configuration, flows, refactoring, Apex, integrations, triggers, Lightning web components, testing, and documentation. Whether you are grappling with Lightning web components or orchestrating intricate integrations, you can jump directly to the chapter that resonates with your immediate needs.

However, if you're newly embarking on this exciting journey, we recommend starting with this chapter. This foundational chapter is designed to set you up for success by detailing the process of configuring ChatGPT specifically for Salesforce development. It lays the groundwork and provides the essential insights that will enable you to make the most out of the following chapters.

The beauty of this book lies in its adaptability. Whether you're a Salesforce technical architect, a developer, or someone interested in exploring the capabilities of generative AI in transforming Salesforce project delivery, this book offers a rich blend of technical understanding and practical insight. By constructing it as a reference guide, we've ensured that the wealth of information contained within is not just comprehensive but also immediately accessible, reflecting the very nature of generative AI – innovative, adaptable, and practical.

In purchasing this book, you are opening a door to a collaborative experience where your Salesforce projects can be accelerated and enriched by the nuanced application of ChatGPT. The convergence of these two powerful technologies promises an exciting landscape, and this book is your companion, guiding you through every facet of this intricate yet rewarding path.

A little bit about Salesforce

Salesforce, founded in 1999 by Marc Benioff and Parker Harris, revolutionized the software industry by pioneering cloud-based **Software-as-a-Service (SaaS)** with its **customer relationship management (CRM)** platform. In its infancy, Salesforce provided sales automation tools that offered a stark contrast to traditional on-premises solutions. By offering its tools over the cloud, Salesforce lowered barriers to entry, enabling businesses of all sizes to leverage sophisticated CRM functionalities without extensive capital investment.

Over time, the platform evolved from having a primary focus on sales automation into a comprehensive ecosystem encompassing marketing, service, commerce, and more. The introduction of the Force.com platform allowed developers to create custom applications, extending Salesforce's functionality beyond CRM. With powerful tools such as Apex for custom programming, options for custom user interfaces, and a wide array of APIs for integration, Salesforce opened doors for robust customization and configurations, catering to a variety of business needs.

The launch of the Lightning Experience heralded a significant shift in the way developers interact with the platform and introduced Lightning Web Components, which made development more aligned with modern web standards. Simultaneously, Salesforce continued to enhance its declarative tools, empowering administrators and citizen developers to configure complex functionalities without writing code. Today, Salesforce's rich array of features and tools, coupled with a thriving AppExchange marketplace, enables organizations to craft tailor-made solutions that align with their unique business goals, solidifying Salesforce's position as a leader in the cloud computing landscape.

History of AI

The history of generative AI can be traced back to the early thoughts of Alan Turing in the 1950s. Turing, widely recognized as the father of theoretical computer science, proposed the idea of a universal machine that could simulate any human intellect, laying the groundwork for AI. However, it wasn't until the late 20th century that these theoretical ideas began to take concrete form. Douglas Hofstadter's works, notably his 1979 book *Gödel, Escher, Bach*, delved into self-referential systems and the ability of machines to mimic human-like thinking, providing significant insights into generative AI.

Across much of the late 20th century, the development of AI was sidetracked by the symbolic approach, which focuses on rules and logic rather than learning from data. This, combined with high expectations and subsequent disillusionments, led to what is often referred to as "AI winters," periods of reduced funding and interest in AI research. However, during this time, some researchers were planting the seeds for a revival.

A significant shift occurred with the advent of deep learning, spearheaded by researchers such as Geoff Hinton. By embracing neural networks and algorithms capable of learning from vast amounts of data, deep learning brought AI closer to its original goal of emulating human intelligence. These advancements breathed new life into generative AI, giving rise to powerful models capable of creative tasks, such as generating text, music, and images.

The last decade has seen a remarkable surge in AI adoption, with generative AI models becoming increasingly sophisticated and accessible. The development might be best described through Ernest Hemingway's quote from *The Sun Also Rises*: "How did you go bankrupt?" Bill asked. "Two ways," Mike said. "Gradually and then suddenly." This encapsulates the trajectory of generative AI, a field that seemed to progress incrementally and then, bolstered by technological advancements and an explosion of available data, suddenly blossomed into a thriving, transformative force in various industries.

Today, generative AI is not merely a concept; it's a reality driving innovation, personalization, and efficiency across numerous domains. From chatbots to content creation, generative AI has opened doors to new possibilities, allowing businesses and individuals to harness the power of AI in unprecedented ways. It's a testament to the relentless pursuit of a vision that began with Turing's theoretical musings, navigated through periods of stagnation and renaissance, and culminated in a technological revolution that continues to unfold.

Salesforce and generative AI

The intersection of Salesforce, the world's leading CRM platform, and generative AI, a thriving subset of artificial intelligence, represents a shift in how Salesforce projects are delivered and conceived. Salesforce has evolved from a pioneering SaaS model in the late 1990s to a comprehensive suite of solutions that empower businesses to connect with their customers in innovative ways. It offers unprecedented customization and scalability, allowing for tailored solutions that align closely with business goals and user needs.

Generative AI, with its deep roots in mimicking human-like intelligence and creativity, adds an entirely new dimension to Salesforce's capabilities. By incorporating generative AI models such as ChatGPT into Salesforce projects, development teams can automate and accelerate various aspects of the development life cycle. This includes tasks such as configuration, writing and refactoring Apex code, crafting Lightning Web Components, and even generating test scripts. It amplifies efficiency while enabling a deeper and more nuanced understanding of customer requirements and pain points.

What sets this union of Salesforce and generative AI apart is the potential it unlocks in customization and personalization. As Salesforce continues to evolve its offerings, from straightforward CRM functionalities to advanced analytics and integration possibilities, generative AI ensures that the development process keeps pace. The ability to model complex business logic, visualize intricate customer journeys, and even predict future trends through generative models transforms the way Salesforce can be leveraged.

A deep understanding of how to use generative AI tools is becoming an essential asset for development teams in this evolving landscape. The knowledge and skillset required extend beyond mere implementation to strategic alignment with business objectives. Understanding how to harness the power of generative AI within Salesforce's rich ecosystem allows for more innovative solutions that not only fulfill current needs but also anticipate future challenges and opportunities.

The future of Salesforce development, powered by generative AI, looks promising, as it fosters a collaborative environment that is iterative and adaptive. It's not just about delivering projects faster or with fewer resources; it's about reimagining what's possible within the Salesforce platform. The synergy between Salesforce's inherent flexibility and the generative capabilities of AI is poised to redefine how businesses interact with their customers, streamline internal processes, and innovate within their respective industries. It's a confluence that signifies not merely an incremental improvement but a transformative leap in technology-enabled business solutions.

OpenAI and ChatGPT

OpenAI, founded in December 2015, is at the forefront of research and development in AI. With a mission to ensure that **artificial general intelligence** benefits all of humanity, OpenAI focuses on creating and publishing research that pushes the boundaries of AI capabilities. Their commitment to collaboration and openness within the scientific community helps drive rapid advancements and promotes responsible AI development.

ChatGPT is a prime example of OpenAI's vision of generative AI, providing an interface between machine intelligence and human-like text generation. Using the transformative technology of deep learning, ChatGPT is designed to understand and generate human-like responses across diverse contexts and domains. Whether in creative writing, technical problem-solving, or conversational interactions, ChatGPT exemplifies OpenAI's commitment to building safe and beneficial AI. By making such powerful generative models available and customizable, OpenAI not only accelerates innovation across various industries but also continues to explore and address the ethical considerations and societal impacts of widespread AI adoption.

Setting up ChatGPT for Salesforce

Generative AI tools have experienced exponential growth and evolution in recent years. Driven by advancements in deep learning, computational resources, and data availability, these tools are becoming more sophisticated, adaptable, and potent in various applications. The acceleration in their capabilities is remarkable, offering unprecedented opportunities for industries to innovate and solve complex problems. From creating art and music to optimizing business processes and automating coding tasks, the transformational potential of generative AI tools is being realized in diverse domains.

As we delve into the context of Salesforce development, ChatGPT emerges as a valuable asset. This section will guide you through the process of setting up ChatGPT specifically for Salesforce development. From aligning it with Salesforce's ecosystem to tailoring it for crafting solutions, integrations, and test scripts, this comprehensive walkthrough aims to equip you with practical know-how. Whether you're a seasoned Salesforce technical architect or a developer looking to enhance productivity, leveraging ChatGPT for Salesforce development opens doors to innovation, efficiency, and robust solutions that resonate with business goals. The intersection of generative AI with Salesforce's capabilities promises to redefine the paradigms of project execution and delivery.

It's imperative to acknowledge that the directions and insights provided in this book are tailored to ChatGPT as of Fall 2023. Given the rapid pace of technological innovation in the field of generative AI, tools such as ChatGPT are continually evolving, with updates and improvements that may alter certain functionalities or processes. Therefore, you may need to adapt the concepts, methodologies, and instructions described in this book so that they align with the specific version of ChatGPT you are using. Embracing a flexible mindset that accommodates change and staying abreast of updates to the platform will ensure that the principles in this book continue to serve as valuable guidance for accelerating Salesforce projects, irrespective of future advancements in ChatGPT technology.

There are three parts to setting up ChatGPT for a Salesforce project:

1. Creating a dedicated chat thread for the project
2. Filling out the custom instructions
3. Prompting ChatGPT with a creative brief

Let's take a closer look.

Dedicated chat thread

ChatGPT can retain context within chat threads, making it an intelligent tool for complex conversations. This retention of context means that ChatGPT can remember the flow of dialogue, including prior prompts and responses within a specific thread. By doing so, the model establishes a continuity that mirrors human conversation, allowing for nuanced interactions that build upon previous exchanges. This can be invaluable in scenarios such as Salesforce development, where a series of related queries may lead to intricate, multi-step solutions. Contextual awareness helps in crafting responses that align with the evolving nature of a conversation, making the interaction more coherent and tailored to the user's ongoing needs and inquiries.

> **Note**
> Click on the pencil icon to the right of the chat name and rename the chat for your project so that it is easy to find and use.

ChatGPT custom instructions

ChatGPT's custom instructions are a sophisticated feature that allows users to guide the model's behavior more explicitly, tailoring its responses to specific needs and contexts. By using custom instructions, a user can set parameters or conditions that the model will adhere to in its replies, ensuring that the content generated is aligned with particular requirements, such as tone, style, or subject matter expertise.

For example, in a Salesforce development environment, custom instructions can be employed to ensure that the model's responses are consistent with the technical complexity and specific terminologies relevant to the project. This can significantly enhance efficiency as the generated content will be more immediately applicable without extensive adjustments.

The value of using custom instructions with ChatGPT is found in the ability to create a more controlled and targeted interaction with the model. It bridges the gap between a general-purpose AI and a specialized tool, enabling the user to customize responses according to the specific demands of their projects. Whether in creative writing, technical documentation, or intricate problem-solving, custom instructions empower users to leverage the power of generative AI in a way that's highly relevant and tailored to their unique needs and goals.

The custom instructions should be fine-tuned to each project and the individual's role in the project. A good starting point is provided here:

What would you like ChatGPT to know about you so that it can provide better responses?	I am a Salesforce technical architect and developer with the IIBA Certified Business Analysis Professional certification. My background exemplifies a rare and valuable blend of skills in the modern IT landscape. I have decades of experience in the Salesforce ecosystem, which gives me unparalleled expertise in crafting solutions that align with business goals, leveraging the rich features and capabilities of Salesforce's suite of products.
	My technical acumen in Salesforce includes in-depth knowledge of AppExchange, Apex, Lightning Web Components, and integrations with various third-party systems. The ability to architect scalable and robust solutions, all while considering security, performance, and compliance requirements, sets me apart from my peers.
	Furthermore, the possession of the IIBA Certified Business Analysis Professional certification showcases my adeptness in identifying, defining, and analyzing business needs. This includes translating those needs into functional and non-functional requirements and collaborating with stakeholders to ensure alignment with strategic objectives.
	My decades of hands-on experience in Salesforce project management and delivery have allowed me to learn and fine-tune methodologies that enable successful project execution, balancing scope, time, cost, and quality. My approach to collaboration and innovation ensures the delivery of tailored solutions and also contributes to driving organizational success and growth.

How would you like ChatGPT to respond?	The ChatGPT service should respond to prompts like it is a seasoned Salesforce technical architect and developer and must demonstrate a rich blend of technical understanding and practical insight. Considering the user is a hands-on developer working on Salesforce configuration and coding based on user stories, the response must use language that is technical yet accessible.
	The service must possess the capability to interpret complex prompts and translate them into actionable insights and guidelines, providing solutions that are not only innovative but also applicable in the real-world context of Salesforce development.
	The tone of the response should be authoritative, reflecting the wisdom drawn from decades of experience, yet devoid of any condescension. The interaction must be akin to a mentor and a peer sharing insights, where the underlying theme is collaboration and mutual growth.
	ChatGPT's responsiveness to prompts must embody a synthesis of technological acumen, real-world applicability, and a respectful and engaging tone. This will create an environment that promotes exploration, learning, and the application of ChatGPT in transforming Salesforce project delivery, thereby aligning with the vision of accelerating everything people do with the power of generative AI.

Salesforce project creative brief

Starting a ChatGPT chat thread for a Salesforce project with a creative brief serves as a strategic alignment tool, allowing generative AI to be integrated within the specific contours of the project's goals, requirements, and constraints. By defining the essential aspects of the project within the brief, custom instructions can be precisely crafted to reflect the unique nature of the Salesforce implementation, focusing on areas such as configuration, Apex coding, Lightning Web Components, and testing.

This approach creates a symbiotic relationship between the detailed planning inherent in the creative brief and the customizable features of ChatGPT, ensuring that the AI's output matches the Salesforce project's complexity and specificity. It means that the developers, architects, and business analysts involved in the project can collaborate with ChatGPT in a more nuanced and insightful manner, with the AI being sensitive to the particular Salesforce context and user role, including aspects such as security, performance, compliance, and alignment with business goals.

In essence, using a creative brief to refine custom instructions for a Salesforce project is akin to tuning a finely crafted instrument. It ensures that the power of generative AI is harnessed in a way that's not only technically adept but also creatively inspired, facilitating an agile and innovative development process. It sets the stage for ChatGPT to function as a responsive and insightful partner in the Salesforce development journey, contributing to tailored solutions that can drive organizational success and growth.

The creative brief should be the first prompt in the chat thread for the new project:

Creative brief	This is the creative brief for the project you're going to help me with. There is no need to respond to this prompt, though I am looking forward to your answers to the subsequent prompts.
	Accelerating Salesforce: Leveraging generative AI for enhanced configuration and customization
	Objective: To provide authoritative and accessible guidance and support for a Salesforce developer, focusing on the utilization of generative AI to accelerate and improve the customization and configuration of Salesforce for my project.
	Target audience: Salesforce developer.
	Tone and style: Authoritative without being condescending, engaging, encouraging, and technical, yet comprehensible to a broad spectrum of Salesforce professionals.
	Call to action: Encouraging continuous learning and exploration of both Salesforce and generative AI while guiding toward online forums, resources, and communities.
	Conclusion: This project aims to empower Salesforce users to innovate and excel in their use of Salesforce.

ChatGPT-assisted Salesforce development

Utilizing ChatGPT for a Salesforce project is akin to engaging in a collaborative partnership, one that transcends the conventional approach of simply extracting code samples or isolated insights. In this partnership, ChatGPT plays the role of an astute colleague, offering insights, proposing solutions, refining ideas, and contributing to all aspects of the project from design through testing. The relationship is interactive and iterative and is built on continuous dialogue and exchange. Approach this collaboration by doing the following:

- **Initiating the dialogue**: Start by framing your problem or need within the context of your Salesforce project. Approach ChatGPT as a thinking partner, laying out the scenario, the constraints, and the goals, and seeking guidance or suggestions. This initial interaction sets the stage for the dialogue that follows.

- **Iterative exploration**: The beauty of ChatGPT lies in its ability to engage in a nuanced back-and-forth conversation. You don't simply extract an answer and move on. Instead, you probe, challenge, refine, and iterate, treating ChatGPT as a fellow developer who helps you refine your thoughts and crystallize your solutions. This iterative process allows for a rich exploration of possibilities and fosters creative problem-solving.

- **Collaborative development**: As you delve into development, from configuration and coding to integrations and Lightning Web Components, treat ChatGPT as a co-developer. Share your progress, ask for code reviews, seek alternative solutions, and leverage the AI's understanding of Salesforce's rich features. This collaborative approach ensures that you are tapping into the breadth and depth of expertise that ChatGPT offers.

- **Testing and refinement**: ChatGPT's involvement extends to the testing phase. Here, again, the iterative dialogue continues, where you can share test results, explore anomalies, create test scripts, and work together to refine and optimize the solution. The approach is hands-on and guided by real-world applicability and compliance.

- **Avoiding over-reliance on templates**: It's essential to avoid treating ChatGPT as a mere repository of code snippets or templates. While it can provide code examples, the real value lies in its ability to engage in problem-solving, offer tailored solutions, and adapt to the unique needs and constraints of your Salesforce project.

- **Aligning with best practices**: As a partner in development, ChatGPT can also be a guiding force in adhering to best practices, security protocols, and performance standards. Engaging in dialogue about these aspects ensures that your project is not only innovative but also robust and compliant.

Working with ChatGPT on a Salesforce project is a dynamic and rewarding collaboration that's characterized by continuous dialogue, iterative exploration, and shared problem-solving. It's a partnership where you harness the full potential of generative AI, not by merely copying and pasting solutions, but by actively engaging, probing, and refining until the best possible solution is realized. It is, in many ways, a new paradigm of development, where the boundaries between human insight and AI capabilities blend into a seamless and highly effective development journey.

Troubleshooting with ChatGPT

ChatGPT ingests vast amounts of data through an initial training process, encompassing everything from text and code to technical documents and contextual narratives. This includes specific details about platforms such as Salesforce, capturing nuances of functionalities, errors, resolutions, and best practices. Unlike a human, who learns through sequential experience and study, ChatGPT's ingestion process assimilates and synthesizes a broad spectrum of information in a highly efficient manner.

The data that's ingested by ChatGPT is stored in a highly complex mathematical representation within neural networks. These networks encode relationships, patterns, context, and interdependencies, preserving them in a manner that's orders of magnitude more extensive and precise than human memory. This allows ChatGPT to maintain perfect recall of the vast information landscape to which it has been exposed.

When responding to a prompt, ChatGPT's neural networks process the input, navigating the intricate web of information stored within. By recognizing patterns, understanding context, and applying logic, ChatGPT crafts responses that are not only accurate but also contextually relevant. This capability is especially potent when applied to complex ecosystems such as Salesforce.

Salesforce is a rich and multifaceted platform that continues to evolve. Keeping up with every facet of Salesforce is a daunting task for any individual. ChatGPT's unique advantage lies in its ability to have "ingested" knowledge from myriad sources, integrating insights, case studies, solutions, and documentation. This results in a composite understanding that transcends what any single human expert could achieve.

It just takes ChatGPT ingesting one instance of a Salesforce error and its resolution for it to become part of ChatGPT's vast repertoire. Unlike human knowledge, which may remain confined to an individual or team, ChatGPT democratizes this knowledge, making it available for future troubleshooting. This promotes efficiency and standardizes the problem-solving approach across diverse scenarios.

Human memory is fallible and constrained by capacity. ChatGPT's perfect recall means that it can quickly retrieve relevant information, be it an obscure Salesforce configuration detail, a nuanced coding solution, or a new feature published publicly for the first time just hours earlier. This instant access to precise information sets ChatGPT apart as an extraordinary resource.

As a Salesforce troubleshooting partner, ChatGPT offers a blend of extensive knowledge, perfect recall, and adaptability. It can guide developers through complex error landscapes, propose solutions, validate approaches, and engage in iterative problem-solving. The partnership is dynamic, responsive, and built on the foundation of collective intelligence.

While ChatGPT's capabilities are immense, it doesn't replace human insight, intuition, and contextual understanding. What it does offer is a complementary resource that amplifies human expertise. In troubleshooting Salesforce, the synergy between human judgment and ChatGPT's vast knowledge pool creates a robust problem-solving environment.

As Salesforce continues to evolve, so does the need for a dynamic and adaptable knowledge base. ChatGPT's architecture enables continuous learning and adaptation, ensuring that it remains aligned with the evolving landscape of Salesforce, its features, challenges, and solutions.

In the context of Salesforce, ChatGPT offers a new paradigm of collaboration and expertise. It's not a mere tool or repository but an active, engaging, and invaluable partner in navigating the complexities of Salesforce. Its vast knowledge, perfect recall, and dynamic interaction make it an essential asset for anyone working within the Salesforce ecosystem, fostering efficiency, innovation, and excellence.

ChatGPT "levels up" developers

In the Salesforce ecosystem, the demarcation between Salesforce administrators and developers is increasingly blurred, primarily due to the platform's versatile and user-friendly nature. Salesforce, with its rich suite of declarative tools and features, empowers administrators to perform tasks that have traditionally fallen in the developer's domain. This transition is accentuated by the platform's design, encouraging a low-code or no-code approach, enabling administrators to implement complex business logic and automation without delving into the intricacies of coding (until they are ready to do so). As a result, administrators often find themselves easing into roles that require a blend of both administrative and developmental skills. They start by tweaking existing processes and gradually move toward building more complex solutions, bridging the gap between administrative tasks and custom development. This trend reflects Salesforce's vision of making technology accessible and empowering users to tailor solutions closely aligned with business objectives, thereby fostering a seamless transition from administration to development.

Integrating tools such as ChatGPT into the development process is revolutionizing the way developers acquire and apply knowledge. By making extensive, real-world expertise accessible on demand, it effectively "levels up" the skills of developers, narrowing the gap between those with only a few years of experience and seasoned veterans.

For a developer with 1 to 2 years of experience, traditional paths to acquiring the deep knowledge that comes with 5 or more years of hands-on work might include extensive reading, attending workshops, or seeking mentorship from more experienced colleagues. While these methods are valuable, they are time-consuming and may not always be readily accessible. ChatGPT provides immediate access to insights that might otherwise take years to accumulate, delivering specific, tailored information in mere moments.

Unlike generic online tutorials or forums, ChatGPT offers personalized guidance that aligns with the unique context and needs of a developer's project. Whether it's troubleshooting a specific issue or designing a complex architecture, the information is tailored to the developer's current problem, enabling them to proceed with confidence and efficiency. This bespoke assistance simulates the mentorship that a seasoned developer might provide, but with the speed and availability of a digital tool.

Continuous learning is vital in the rapidly evolving field of software development. By engaging with ChatGPT, developers not only get answers to immediate questions but also expose themselves to broader concepts, best practices, and innovative approaches. This accelerated learning curve supports rapid growth, allowing a developer with less experience to operate at a level that might otherwise require many more years of practice and study.

Collaboration is key to successful development, and ChatGPT acts as a virtual team member, contributing ideas and solutions. While it doesn't replace human collaboration, it enhances it by offering a resource that can be consulted at any time, fostering a collaborative problem-solving environment. This collaboration can lead to more creative and effective solutions, mirroring the synergy that often exists within highly experienced development teams.

In organizations where there is a diverse range of experience levels, ChatGPT can act as a leveling tool, ensuring that all developers, regardless of their tenure, have access to the same high-quality information and guidance. This can foster a more cohesive and agile team, where junior members can contribute at a higher level, and senior members can focus on more complex challenges.

ChatGPT represents a paradigm shift in how developers acquire and apply knowledge. By offering instant access to the wisdom and insights typically gained through years of experience, it empowers developers at all levels to work more effectively and innovatively. The knowledge gap that traditionally separated junior and senior developers can be bridged in a matter of minutes, democratizing access to expertise and fostering a more dynamic and collaborative development landscape.

The case study

Some of the chapters in this book make use of the case study in *Appendix A*.

Acme Corp, having recently deployed Salesforce Sales Cloud with an integration to Oracle ERP, is seeking to extend its capabilities by integrating Snowflake, which is already in use for AI-based analytics. The business leadership is exploring a solution that would include a "Snowflake" section on the Salesforce Account pages to summarize information and enable sales reps to select related contacts and automatically copy the Snowflake information to emails.

This case study, which starts with a business problem statement, demonstrates the use of ChatGPT to go from a business problem statement to epics to user stories. The user stories will be used throughout this book.

Summary

This chapter opened the door to integrating ChatGPT with Salesforce, demonstrating how conversational AI can revitalize the development process on the platform. It set the stage for a deep dive into using ChatGPT to streamline Salesforce projects. This chapter not only introduced the foundational concepts but also imparted essential practices to effectively harness ChatGPT in the Salesforce environment. It underscored the necessity of aligning with best practices and showcased a case study that you will follow throughout this book, highlighting the practical benefits and learning opportunities of ChatGPT integration. This chapter promised a transformative journey, equipping developers with knowledge on seamless integration, efficient development strategies, troubleshooting techniques, continuous learning habits, and adherence to best practices – all critical skills in the modern Salesforce development landscape.

In the next chapter, you'll learn how to use ChatGPT for Salesforce configuration. Configuration is often a balancing act between the capabilities of the platform and the specific needs of the business, and in this chapter, you will learn how to walk this tightrope with ChatGPT as your safety net. You will gain a clear understanding of how configuration differs fundamentally from coding and how ChatGPT can provide valuable insights that streamline this process. By the end of the next chapter, you will have developed a robust understanding of the Salesforce configuration landscape. You will have learned how to articulate queries to ChatGPT, translate its advice into practice, and troubleshoot with efficiency. These skills will empower you to tackle configuration tasks with confidence while using ChatGPT's AI-driven insights to optimize Salesforce to your organization's unique demands.

2

Using ChatGPT
for Salesforce Configuration

Salesforce, since its inception in the late 1990s, has been guided by a unique vision: "clicks, not code." This philosophy has been at the core of their business model and product development strategy, reflecting a conscious effort to provide solutions that address most business needs through configuration rather than customization.

The underlying idea of "clicks, not code" is to empower businesses, even those with minimal technical expertise, to harness the power of Salesforce's suite of products without having to write custom code. Through its various declarative tools, Salesforce allows users to define process flows, create custom objects, set validation rules, and manage permissions, all through a user-friendly interface. This approach opens up opportunities for a broader range of individuals within an organization, such as administrators and business analysts, to actively participate in the configuration and management of the system.

Over the past two and a half decades, Salesforce has heavily invested in building features that align with this vision. They have expanded their platform to encompass various aspects of customer relationship management (CRM), sales, marketing, and more, all while focusing on tools that enable end users to tailor the system to their needs without writing code. The extensive AppExchange marketplace is a testament to this approach, offering a wide array of plug-and-play solutions that further reduce the need for custom development.

However, it's essential to recognize that Salesforce's dedication to "clicks, not code" doesn't entirely negate the role of custom coding. More complex requirements and integrations might still necessitate the involvement of developers and the use of languages such as Apex and Lightning Web Components. Yet, this coexistence of declarative and programmatic tools within the Salesforce ecosystem speaks to a nuanced approach, where the primary emphasis remains on enabling users to achieve as much as possible through configuration.

By continuously nurturing this philosophy, Salesforce has created a versatile platform that can cater to a diverse set of business requirements, ranging from simple configurations to sophisticated customizations. The ongoing commitment to "clicks, not code" stands as a symbol of Salesforce's dedication to user empowerment and innovation, paving the way for more inclusive, efficient, and flexible solutions.

The basics of Salesforce configuration

Across the Salesforce ecosystem, there are tens of thousands of possible configuration changes. This chapter will not attempt to explain all of these changes. Rather, the intent is to show the different ways ChatGPT can be used to understand and make configuration changes. All configuration changes will be addressed through Salesforce admin screens, with a smaller number also performed by editing Salesforce XML.

Here is a list of five common configuration changes that administrators and developers often make to Salesforce to tailor the platform according to specific business needs:

- **Creating custom fields and objects**: To capture unique data requirements, administrators and developers create custom fields within standard objects or even design entirely new custom objects.

- **Setting up validation rules**: Validation rules are used to ensure data integrity by defining specific criteria that must be met before a record can be saved.

- **Modifying page layouts**: Modifying page layouts helps in presenting the most relevant information to different user profiles in a way that aligns with their roles and responsibilities.

- **Creating a report**: Creating custom reports provides tailored insights and analytics that match the specific needs and KPIs of the organization.

- **Setting up approval processes**: Approval processes are used to automate the routing of records through various approval steps, mirroring a company's specific approval hierarchy and rules.

These common configuration changes demonstrate the flexibility of the Salesforce platform and its ability to be updated to meet diverse and evolving business requirements. The combination of these functionalities enables organizations to craft a CRM system that aligns closely with their unique processes and objectives.

User stories serve as a critical bridge in translating business requirements into technical solutions, especially within the context of Salesforce configuration. They encapsulate the needs and desires of the end-users in a concise and understandable format, often structured around the perspective of a specific user role and what they hope to accomplish. This standardized communication tool enables a clear and focused dialogue between business stakeholders and developers. By defining the "who," "what," and "why" of a requirement, user stories lay the groundwork for creating configuration changes that align with the business's goals. In this chapter, we will utilize user stories as the basis for defining and explaining various configuration examples, leveraging them as a practical framework that resonates with both the technical and non-technical audience, and ensuring that the configuration changes are centered on delivering real value to the users.

Creating custom fields and objects

Salesforce objects and fields are fundamental building blocks within the Salesforce platform, forming the core structure for storing and organizing data. Objects in Salesforce are akin to tables in a database; they serve as containers for data and are categorized into two types: standard objects and custom objects. Standard objects are pre-defined by Salesforce, such as Accounts, Contacts, or Opportunities, and are common across most Salesforce implementations, reflecting common business entities. Custom objects, on the other hand, are user-defined and tailored to specific business needs, allowing organizations to store information unique to their processes. Within these objects, fields represent individual data points, similar to columns in a database table. Fields can hold various types of data, such as text, numbers, dates, or even relationships to other objects (lookup or master-detail relationships). This structure enables users to capture a wide range of business-specific information, making Salesforce a highly customizable and flexible platform for managing diverse business processes and relationships. The meticulous design of objects and fields in Salesforce allows for detailed data modeling, which is essential for effective CRM and business process management.

Salesforce custom fields are specialized fields that are added to objects within the Salesforce platform to capture unique business-specific information. Unlike standard fields that come predefined within Salesforce objects (such as Account, Contact, and Opportunity), custom fields are created by administrators or developers to tailor the system according to specific organizational needs and processes. These fields can be of various data types, such as text, number, date, or even references to other objects (lookup fields). The addition of custom fields enables organizations to enhance their Salesforce environment, providing detailed, relevant data that standard fields may not accommodate, thus enriching the quality and context of the data within the system. Custom fields are central to creating a Salesforce setup that aligns with the unique characteristics and requirements of a business, facilitating more targeted actions, reporting, and decision-making.

Let's consider a user story for adding two custom fields to the Salesforce Contact object. The first field is "VIP." VIP stands for Very Important Person and is a checkbox that's used to indicate to sales reps that their organization holds this contact in high regard. The second field is "Relationship Manager." When populated, this indicates that before the sales rep interacts with this contact, they need to get permission from the Relationship Manager. The "Relationship Manager" field is a Salesforce Lookup field to the User object.

User story – add VIP and Relationship Manager fields to Salesforce Contacts:

- **Goal**: As a sales rep, I want to know if a contact is a VIP to my organization and if my organization has appointed a Relationship Manager for the contact.

- **Acceptance criteria**:

  - **Given** a sales rep has navigated to a Salesforce Contact page
  - **When** the page is loaded
  - **Then** the sales rep can view **VIP**, which is a checkbox field
  - **And** the sales rep can view **Relationship Manager**, which is a lookup to the User object

The first approach to using ChatGPT to address this user story is to write a prompt asking ChatGPT to write directions on how to create these fields using the Salesforce admin screens. As noted in *Chapter 1*, ChatGPT is an assistant, and writing prompts is an iterative process. Let's look at the first pass at writing a request to have ChatGPT explain how to address the user story via admin screens.

Please provide detailed step-by-step directions to implement the following user story using Salesforce admin screens.

Combine the two to create a ChatGPT prompt:

Prompt	Please provide detailed step-by-step directions to implement the following user story using Salesforce admin screens.
	
	 User story – Add VIP and Relationship Manager fields to Salesforce Contacts:
	
	 • Goal: As a sales rep, I want to know if a contact is a VIP to my organization and if my organization has appointed a Relationship Manager for the contact.
	
	 • Acceptance criteria:
	
	     ▪ **Given** a sales rep has navigated to a Salesforce contact page
	
	     ▪ **When** the page is loaded
	
	     ▪ **Then** the sales rep can view **VIP**, which is a checkbox field
	
	     ▪ **And** the sales rep can view **Relationship Manager**, which is a lookup to the User object

Response	Certainly! Here are the step-by-step instructions to implement the given user story by adding the "VIP" and "Relationship Manager" fields to Salesforce Contacts through the Salesforce admin screens: 1. Log into Salesforce: Navigate to your Salesforce instance and log in using the credentials that provide administrator access. 2. Access Object Manager: Click on the setup gear icon (⚙) in the top-right corner and then click on Object Manager.

3. Navigate to the Contact object: In Object Manager, search for and click on Contact.

4. Access fields and relationships: Click on Fields & Relationships in the left-hand navigation pane.

5. Create the VIP checkbox field:

  - Click New.

  - Select Checkbox as the data type and click Next.

  - Label the field VIP and provide a description if desired.

  - Set Default Value to unchecked (or checked if you prefer).

  - Click Next.

  - Select the profiles that need access to this field, and set the visibility permissions.

  - Click Next.

  - Add the field to the appropriate page layouts by checking the boxes next to the layout names.

  - Click Save.

6. Create the Relationship Manager lookup field:

  - Click New.

  - Select Lookup Relationship as the data type and click Next.

  - Choose User from the Related To drop-down menu.

  - Label the field Relationship Manager and provide a description if desired.

  - Click Next.

  - Select the profiles that need access to this field, and set the visibility permissions.

  - Click Next.

  - Add the field to the appropriate page layouts by checking the boxes next to the layout names.

  - Click Save.

	7. Verify the changes:
	- Navigate to any Contact record within Salesforce.
	- Verify that the VIP checkbox field and Relationship Manager lookup field are visible and functioning as per the acceptance criteria.
	By following these steps, the two new fields will be successfully added to the Salesforce Contact object, fulfilling the user story's requirements. The sales rep can now easily identify a VIP contact and view the assigned Relationship Manager.

The provided user story and the corresponding directions leave several essential details unspecified, which could lead to ambiguity during the implementation phase. The user story doesn't define which specific profiles or permission sets should have read and update permissions for the new **VIP** and **Relationship Manager** fields, leaving a gap in the security and accessibility controls. Similarly, the details about the exact page layouts to which the fields should be added are missing, and this may result in inconsistency across different layouts within Salesforce. Furthermore, the desired placement of these fields on the page layouts is not described, leading to potential confusion about the intended user experience and layout design. These gaps highlight the importance of a more detailed specification in the user story, acceptance criteria, and implementation instructions to ensure alignment with the business needs and to facilitate a smooth and efficient configuration process. *Chapter 11* goes into detail about how to use ChatGPT to groom user stories and what questions to ask the business analyst and product owner during the grooming process.

For this chapter, however, we'll use a user story as the basis for a ChatGPT prompt. With this prompt, ChatGPT has responded with detailed step-by-step configuration directions!

For experienced developers, these updates can also be made directly in the Salesforce org XML:

> **Note**
> Ensure you know what you are doing when updating Salesforce XML directly! While this is faster than navigating through multiple admin screens, it also lacks many of the guardrails provided by the admin screens.

Prompt	Please rewrite your response to the previous prompt to provide the Salesforce metadata XML for these two contact custom fields, the profile updates, and the Lightning page layout updates so that the XML can be pasted into a Salesforce org using a tool such as Visual Studio Code.

| Response | Certainly! Here are the Salesforce metadata XML definitions for the two new custom fields on the Contact object, profile updates, and Lightning page layout updates. You can paste this XML code into your Salesforce organization using a tool such as Visual Studio Code. |

Custom fields:

```
<CustomObject xmlns="http://soap.sforce.com/2006/04/
metadata">
    <fields>
        <fullName>VIP_Customer__c</fullName>
        <description>Indicates if the contact is a VIP
customer</description>
        <inlineHelpText>Check this box if the contact
is a Very Important Person (VIP) to the organization.</
inlineHelpText>
        <label>VIP Customer</label>
        <type>Checkbox</type>
        <defaultValue>false</defaultValue>
    </fields>
    <fields>
        <fullName>Relationship_Manager__c</fullName>
        <description>Reference to the Salesforce User who
is the Relationship Manager for the contact</description>
        <inlineHelpText>Select the Relationship Manager
for this contact from the User object.</inlineHelpText>
        <label>Relationship Manager</label>
        <type>Lookup</type>
        <referenceTo>User</referenceTo>
        <relationshipName>Relationship_Manager</
relationshipName>
    </fields>
</CustomObject>
```

Profile updates:

```
<Profile xmlns="http://soap.sforce.com/2006/04/metadata">
    <fieldPermissions>
        <editable>true</editable>
        <field>Contact.VIP_Customer__c</field>
        <readable>true</readable>
    </fieldPermissions>
    <fieldPermissions>
```

```
            <editable>true</editable>
            <field>Contact.Relationship_Manager__c</field>
            <readable>true</readable>
        </fieldPermissions>
        <!-- Add other permissions and settings as needed -->
    </Profile>
```

Lightning page layout updates:

For the Lightning page layout, you may want to use tools such as Lightning App Builder to visually configure the page. Here is an example representation of the XML configuration. Please note that the exact XML will depend on your specific Lightning page layout:

```
<flexiPage xmlns="http://soap.sforce.com/2006/04/
metadata">
    <flexiPageRegions>
        <componentInstances>
            <componentName>force:detailPanel</
componentName>
        </componentInstances>
        <name>main</name>
        <type>Region</type>
    </flexiPageRegions>
    <masterLabel>Contact Page with VIP and Relationship
Manager</masterLabel>
    <!-- Additional configurations specific to your
Lightning layout -->
</flexiPage>
```

This XML is illustrative and might require specific adjustments based on the existing Lightning page layout in your Salesforce organization. The Lightning page layout often involves more intricate configurations that might best be handled through the Lightning App Builder interface, from which the XML metadata can be extracted.

Please make sure you adapt this XML to your specific organization's needs, including naming conventions, and reference the correct profiles and layouts. If multiple profiles or layouts are involved, repeat and customize the appropriate sections as needed.

In addressing the user story, the guidance provided by ChatGPT caters to developers of varying skill levels, creating a bridge between the novice and the experienced. For those who are just getting started with Salesforce development, the response initially provided a clear step-by-step process that navigated the Salesforce admin screens. This approach was designed to make the implementation as accessible as possible for newcomers. Conversely, for the more seasoned developers who have a "hardcore gets under the covers" approach, the guidance shifted to delivering the Salesforce metadata XML code. This format allowed for more nuanced and customized control over the configuration, catering to those who prefer working directly with metadata and utilizing tools such as Visual Studio Code. By providing these different methodologies, ChatGPT demonstrates the capacity to offer solutions that align with both the simplicity sought by newcomers and the intricacy desired by seasoned professionals, nurturing growth and efficiency in the Salesforce development process.

One note about the ChatGPT response – while it is correct and will work, it does not reflect the latest approach recommended by Salesforce. While ChatGPT accelerates Salesforce delivery, developers who know their way around Salesforce are still needed!

Setting up validation rules

Salesforce validation rules play a crucial role in maintaining the integrity and accuracy of data within the Salesforce environment. These rules are set up to enforce specific criteria and conditions that must be met before a record can be saved. By defining expressions and formulas that evaluate the data being entered or modified, validation rules ensure that they adhere to the established business requirements and logic. For example, a validation rule might require that a discount percentage field on an opportunity record must fall within a specific range, or that a mandatory phone number field must follow a specific format. If the conditions are not met, a user-friendly error message can be displayed, guiding the user to correct the input. Through this mechanism, validation rules act as gatekeepers, promoting data quality, consistency, and compliance with organizational standards and processes.

Continuing with the user story that was introduced in the first example, the next user story requires a Relationship Manager to be selected when the contact is a VIP.

User story – *require a Relationship Manager when the contact is a VIP*:

- **Goal**: As a sales rep, I want contacts set as VIP to my organization to always have a Relationship Manager

- **Acceptance criteria**:

  - **Given** a sales rep has navigated to a Salesforce contact page

  - **When** the sales rep checks the **VIP** checkbox

  - **Then** the **Relationship Manager** field can't be blank

We'll use the same request – that is, Please provide detailed step-by-step directions to implement the following user story using Salesforce admin screens.

Let's build a ChatGPT prompt:

Prompt	Please provide detailed step-by-step directions to implement the following user story using Salesforce admin screens.
	User story – require a Relationship Manager when the contact is a VIP:
	• **Goal**: As a sales rep, I want contacts set as VIP to my organization to always have a Relationship Manager
	• **Acceptance criteria**:
	▪ **Given** a sales rep has navigated to a Salesforce contact page
	▪ **When** the sales rep checks the **VIP** checkbox
	▪ **Then** the **Relationship Manager** field can't be blank
Response	Certainly! Implementing the user story that requires a Relationship Manager to be assigned when the contact is marked as a VIP can be done through Salesforce validation rules. Here's how to achieve this using Salesforce admin screens:
	1. Log into Salesforce: Sign into the Salesforce instance with administrator credentials.
	2. Navigate to Object Manager: Click on App Launcher (grid icon) and type "Object Manager." Then, click Object Manager.
	3. Select the Contact object: In Object Manager, find and click Contact.
	4. Create a new validation rule: Click on Validation Rules in the left sidebar, then click the New button.
	5. Define the validation rule:
	a. Name the rule: Provide a name for the validation rule, such as Require_Relationship_Manager_for_VIP.
	b. Enter the error condition formula: In the Formula text area, enter AND(ISPICKVAL(VIP__c, true), ISBLANK(Relationship_Manager__c)). Replace VIP__c and Relationship_Manager__c with the appropriate API names for your custom fields.
	c. Set the error message: Define the error message that will be shown if the condition is met, such as the Relationship Manager must be selected for VIP Contacts.
	d. Choose an error location: Decide where the error message will be displayed. You can choose the field that triggered the error or the top of the page.

	6. Save the validation rule: Click the Save button to save the validation rule.
	7. Test the rule: Create or edit a Contact record, try to set it as a VIP without selecting a Relationship Manager, and confirm that the validation rule prevents the record from being saved and displays the appropriate error message.
	This step-by-step process ensures that when the VIP checkbox is checked, the Relationship Manager field must contain a value. If it doesn't, the error message will be displayed, enforcing the required business logic within the Salesforce org.

As with the custom fields example, ChatGPT can produce XML to be pasted directly into the Salesforce org.

Modifying page layouts

Updating Salesforce Lightning pages involves a flexible and dynamic approach that enables organizations to tailor the user interface so that it meets specific business needs. Salesforce administrators or developers can utilize Lightning App Builder, a drag-and-drop tool, to add, remove, or rearrange components on a Lightning page. This allows for easily customizing layouts, integrating standard or custom components, and setting component attributes or visibility rules. Additionally, Lightning pages can be assigned to different apps, record types, or user profiles, ensuring a personalized user experience. Whether you're creating brand new Lightning pages or modifying existing ones, this approach empowers teams to create engaging, efficient interfaces aligned with their organization's processes and objectives.

Continuing to work with the **VIP** and **Relationship Manager** fields, let's look at a user story to update the page layout used by sales reps.

User story – *Add VIP and Relationship Manager fields to a new tab called "VIP Mgmt" on the primary Salesforce Contact page layout used by sales reps*:

- **Goal**: As a sales rep, I want a tab called **VIP Mgmt** on the Contact page layout I use
- **Acceptance criteria**:
  - **Given** a sales rep has navigated to a Salesforce contact page
  - **When** the page is loaded
  - **Then** the sales rep can click on a **VIP Mgmt** tab
  - **And** the sales rep can view **VIP**, which is a checkbox field
  - **And** the sales rep can view **Relationship Manager**, which is a lookup to the User object

We'll use the same request – that is, Please provide detailed step-by-step directions to implement the following user story using Salesforce admin screens.

Let's build a ChatGPT prompt:

Prompt	Please provide detailed step-by-step directions to implement the following user story using Salesforce admin screens.
	User story – Add VIP and Relationship Manager fields to a new tab called "VIP Mgmt" on the primary Salesforce Contact page layout used by sales reps:
	• **Goal:** As a sales rep, I want a tab called **VIP Mgmt** on the Contact page layout I use
	• **Acceptance criteria**:
	▪ **Given** a sales rep has navigated to a Salesforce contact page
	▪ **When** the page is loaded
	▪ **Then** the sales rep can click on the **VIP Mgmt** tab
	▪ **And** the sales rep can view **VIP**, which is a checkbox field
	▪ **And** the sales rep can view **Relationship Manager**, which is a lookup to the User object
Response	Certainly! Here's how you can implement the user story by adding the VIP and Relationship Manager fields to a new tab called VIP Mgmt on the primary Salesforce Contact page layout used by sales reps:
	1. Log into Salesforce: Log into your Salesforce org with an administrator account.
	2. Navigate to Object Manager: From the Setup area, navigate to Object Manager and then click Contact.
	3. Go to Page Layouts: Under the Contact object, click Page Layouts.
	4. Select the appropriate layout: Choose the page layout that's used by the sales reps (for example, Contact Layout).
	5. Edit the layout: Click on the Edit action for the selected layout.
	6. Create a new section: In the page editor, create a new section by dragging the Section component onto the layout. You can label this section VIP Mgmt.
	7. Add the fields:
	a. Drag the VIP checkbox field into the VIP Mgmt section.
	b. Drag the Relationship Manager lookup field (to the User object) into the VIP Mgmt section.

8. Create a new tab: Within the page layout editor, find the tab area and select New to create a new tab. Name this tab VIP Mgmt.

9. Move the section to the tab: Drag the VIP Mgmt section into the VIP Mgmt tab.

10. Adjust permissions: Ensure that the appropriate profiles or permission sets (for example, sales reps) have the necessary read and edit permissions for these fields.

11. Save the layout: Click Save to apply the changes to the layout.

12. Verify the changes: Log in as a sales rep (or use the Login As feature) and navigate to a contact record to verify that the new VIP Mgmt tab appears and that the VIP and Relationship Manager fields are correctly displayed within that tab.

By following these steps, sales reps can now easily manage VIP contacts and their Relationship Managers within a dedicated tab on the Contact page, aligning with the user story's requirements.

Step 10 says "Ensure that the appropriate profiles or permission sets (for example, sales reps) have the necessary read and edit permissions for these fields." ChatGPT can explain how to do this. In the same chat thread, ChatGPT is contextual and "remembers" the previous prompts and responses so that the whole question does not have to be re-asked:

Prompt	Please expand "Step 10. Adjust Permission" and provide step-by-step directions for giving the Sales Rep profile read and edit permissions on the VIP and Relationship Manager fields.
Response	Certainly! Expanding on Step 10, here's a detailed guide to providing the Sales Rep profile with read and edit permissions on the VIP and Relationship Manager fields:
	10. Adjust permissions for the Sales Rep profile:
	a. Navigate to Profiles: From the Setup area, search for Profiles in the Quick Find box and click on Profiles.
	b. Select the Sales Rep profile: Scroll through the list or search for the Sales Rep profile, then click on it to view its details.
	c. Access the Contact object's permissions: In the profile details page, scroll down to the Field-Level Security section and find the Contact object. Click on View next to the Contact object.
	d. Edit field permissions: Click on the Edit button to modify the field permissions.
	e. Find the VIP field: Scroll through the list of fields until you find the VIP custom field.

f. Set read and edit permissions for VIP: Check the boxes in the Read Access and Edit Access columns next to the VIP field.

g. Find the Relationship Manager field: Continue scrolling to find the Relationship Manager custom field.

h. Set read and edit permissions for Relationship Manager: Check the boxes in the Read Access and Edit Access columns next to the Relationship Manager field.

i. Save your changes: Click on the Save button to apply the changes to the profile.

By carefully following these steps, you will ensure that the Sales Rep profile has the required read and edit permissions for both the VIP and Relationship Manager fields, aligning the Salesforce Contact object with the specified requirements of the user story. This grants sales reps the ability to view and modify these fields as needed, facilitating the management of VIP contacts.

Creating a report

Salesforce reports provide an essential means to analyze, visualize, and share data within the organization. These configurable tools allow users to extract, filter, and present data in various formats, such as tables, charts, and dashboards, depending on the business requirements.

Salesforce offers different types of reports, such as Tabular, Summary, Matrix, and Joined, all of which cater to different visualization needs. Users can choose the type that best represents the data they want to analyze. Furthermore, users can create custom reports by selecting specific objects, fields, filters, and criteria, allowing for the segmentation and aggregation of data as per unique business needs.

These reports can also be embedded in dashboards to provide a comprehensive view of key performance indicators (KPIs) and metrics at a glance. They can be scheduled to run at specific intervals and sent to designated stakeholders, ensuring timely dissemination of information. With the appropriate permissions, users can collaborate on reports, sharing insights and annotations. Security settings ensure that sensitive information is accessible only to authorized users.

Salesforce reports offer real-time insights, enabling decision-makers to respond quickly to trends, opportunities, or challenges. Different departments such as Sales, Marketing, Finance, and Customer Support utilize Salesforce reports to track progress, measure success, identify areas for improvement, and make informed decisions.

Salesforce reports translate complex data into comprehensible insights to facilitate the strategic alignment of business operations and goals. Whether for monitoring sales performance, analyzing marketing campaign effectiveness, tracking customer satisfaction, or other business aspects, Salesforce reports play a vital role in empowering organizations to leverage their data intelligently.

An entire book called *Using ChatGPT for Salesforce Reports* could be written on this topic. For this section of this chapter, a prompt will be written to create a report to list the contacts who have the **VIP** checkbox checked and whose mailing address is in California.

Let's look at the user story.

User story – *create a tabular report to list all VIP contacts in California*:

- **Goal**: As a sales rep, I want a report to list all VIP contacts in California

- **Acceptance criteria**:

  - **Given** a sales rep has navigated to the Salesforce **Reports** tab

  - **When** the **VIP Contacts in California** report is run

  - **Then** the sales rep sees a report of all VIP contacts with a mailing address in California

We'll be using the same request – that is, Please provide detailed step-by-step directions to implement the following user story using Salesforce admin screens.

Let's build a ChatGPT prompt:

Prompt	Please provide detailed step-by-step directions to implement the following user story using Salesforce admin screens.
	User story – create a tabular report to list all VIP contacts in California:
	- **Goal**: As a sales rep, I want a report to list all VIP contacts in California
	- **Acceptance criteria**:
	- **Given** a sales rep has navigated to the Salesforce Reports tab
	- **When** the VIP Contacts in California report is run
	- **Then** the sales rep sees a report of all VIP contacts with a mailing address in California
Response	Certainly! Here's a step-by-step guide to creating a tabular report to list all VIP contacts in California for a sales rep within the Salesforce environment:
	1. Navigate to the Reports tab: Log into Salesforce, click on App Launcher (grid icon), and then search for and select Reports.
	2. Create a new report: Click on the New Report button.
	3. Choose the report type: In the Choose Report Type dialogue box, select Contacts under the Standard Report Types section and then click Continue.

4. Define the filters: You will need to add two filters to narrow down the results:

  a. Click on Add Filter.

  b. For the VIP contacts, select the VIP Customer field, choose the equals operator, and set the value to True.

  c. Click on Add Filter again.

  d. For the contacts in California, select the Mailing State/Province field, choose the equals operator, and set the value to California.

5. Select columns: Click on Columns and then select the specific fields you want to display in the report, such as First Name, Last Name, Email, VIP Customer, and Mailing State/Province.

6. Set the format: Since the report is tabular, ensure that the report format is set to Tabular.

7. Run the report: Click the Run button to view the report with the filtered data.

8. Save the report: If the report looks as expected, do the following:

  a. Click on the Save button.

  b. Enter the report name as VIP Contacts in California.

  c. Choose a report folder where you want to save this report for future access.

  d. Click Save again.

9. Share the report: If needed, set the appropriate sharing settings to allow access to other relevant users within the organization.

10. Customize further: If additional customization is needed, such as sorting or grouping, these options can be adjusted under the Outline tab.

Now, the sales rep can navigate to the Reports tab and run the VIP Contacts in California report to see the desired information. The sales rep will see a report of all VIP contacts with a mailing address in California, aligning with the goal and acceptance criteria of the user story.

Setting up approval processes

Salesforce approval processes, while falling under configuration, can be complicated, and many developers are not familiar with their design and setup. This is where ChatGPT shines and can be a time saver.

Prompt ChatGPT to explain Salesforce approval processes and share the best practices around setting up approval processes:

Prompt	What are Salesforce approval processes? How do they work? What are some of the gotchas developers encounter in setting up approval processes? What are the best practices for configuring Salesforce approval processes?
Response	Salesforce approval processes are automated workflows that manage the process of obtaining approval for a particular record within the Salesforce system. They are used to standardize and streamline the way an organization approves items such as discounts, expenses, or any custom object data that requires approval.
	How they work:
	1. Definition: An administrator defines the approval process, specifying the criteria that determine when the record needs approval and who must approve it.
	2. Activation: The process needs to be activated to be utilized.
	3. Submission: Users submit records for approval based on defined criteria.
	4. Approval steps: The process can have multiple steps, with different approvers at each step.
	5. Actions: There are actions for initial submission, approval, rejection, recall, and reassignment, allowing customization of what occurs at each stage.
	6. Notifications: Email notifications can be set up to inform relevant parties about the status of the approval process.
	Gotchas:
	1. Complexity: As approval processes can become complex, it's easy to make mistakes in configuration, leading to issues in the process.
	2. Debugging: Debugging a failed approval process can be challenging as errors might not always be descriptive.
	3. Lack of flexibility: Standard approval processes might not support highly customized or complex business requirements, requiring the use of more sophisticated tools or coding.
	4. Record locking: Records under approval are locked by default, preventing changes. This can cause unexpected problems if not managed properly.

Best practices:

1. Plan ahead: Carefully plan the approval process, considering all business requirements and potential scenarios.

2. Use templates: If applicable, use templates to avoid building processes from scratch.

3. Keep it simple: Avoid unnecessary complexity by using clear criteria and as few steps as possible.

4. Utilize email templates: Customize email notifications to make them informative and actionable.

5. Test thoroughly: Test the approval process in a sandbox or developer environment before deploying to production.

6. Consider record locking: Be aware of record locking and how it might impact other parts of the system.

7. Use Flows: For more complex processes, consider using Flows as they provide more control and flexibility.

8. Monitor and analyze: Regularly monitor the approval processes and analyze how they are working, making necessary adjustments for continuous improvement.

In conclusion, Salesforce approval processes provide a powerful means to control and standardize the approval workflows in an organization. Being aware of potential challenges and following best practices ensures that these processes are robust, efficient, and aligned with the business needs.

While acquiring the knowledge contained in the previous response is indeed attainable through diligent research, sifting through search engine results, or engaging with Salesforce learning materials, the efficiency and speed of ChatGPT's response offer a distinct advantage. In less than 60 seconds, ChatGPT synthesized the essential elements of Salesforce approval processes, including their functionality, common challenges, and best practices. This rapid response time underscores the value of leveraging ChatGPT in a professional context, especially when time is of the essence and quick, authoritative insights are needed. This not only accelerates learning but also frees up valuable time that can be spent on other critical tasks within the Salesforce development lifecycle.

Let's look at the user story for this.

User story – *create an approval process when the Relationship Manager on a contact changes*:

- **Goal**: As a regional sales manager, I want to approve any changes to the Relationship Manager setting for contacts in my territory

- **Acceptance criteria**:

  - **Given** a user has navigated to a Salesforce Contacts page

  - **When** the **Relationship Manager** field is changed

  - **Then** the sales manager for the region has to approve the new Relationship Manager value

  - **And** if the sales manager doesn't approve/deny the new value in one calendar day, the new value is approved

Because approval processes are complex and can have gotchas, in addition to the standard request to provide directions on how to configure the approval request, we will also ask ChatGPT to identify any gaps in the user story that may result in issues.

Please provide detailed step-by-step directions to implement the following user story using Salesforce admin screens. Please note any gotchas or additional information that should be provided to decrease the likelihood of issues during testing or acceptance.

We can combine the enhanced request and the user story to build the ChatGPT prompt:

Prompt	Please provide detailed step-by-step directions to implement the following user story using Salesforce admin screens. Please note any gotchas or additional information that should be provided to decrease the likelihood of issues during testing or acceptance.
	User story – create an approval process when the Relationship Manager on a contact changes:
	• **Goal**: As a regional sales manager, I want to approve any changes to the Relationship Manager setting for contacts in my territory
	• **Acceptance criteria**:
	• **Given** a user has navigated to a Salesforce contacts page
	• **When** the Relationship Manager field is changed
	• **Then** the sales manager for the region has to approve the new Relationship Manager value
	• **And** if the sales manager doesn't approve/deny the new value in one calendar day, the new value is approved

Response	Certainly! Implementing an approval process in Salesforce based on the given user story can be achieved through the following steps, while also noting potential gotchas and areas where additional information might be needed:
	
	1. Define entry criteria:
	
	  - Navigate to Setup | Process Automation | Approval Processes.
	
	  - Select the object (Contact, in this case) for which you are creating the approval process.
	
	  - Define the entry criteria that determine when the approval process is initiated. In this case, you'd likely want to check when the Relationship Manager field has changed.
	
	  - Gotcha: Ensure that the criteria are accurately defined so that the approval process triggers at the correct time.
	
	2. Create approval steps:
	
	  - Define the approval steps that the record must pass through. You may create one step for the regional sales manager's approval.
	
	  - Gotcha: Clarify who the regional sales managers are and whether they are assigned based on territory or other criteria.
	
	3. Set up email templates and notifications:
	
	  - Create or select email templates that will be used to notify the sales manager about the approval request.
	
	  - Gotcha: Without proper notifications, the manager may overlook the request, hindering the process.
	
	4. Configure actions for approvals and rejections:
	
	  - Define what actions should be taken when the request is approved or rejected. This might include updating a field, sending an email, and so on.
	
	5. Configure time-dependent actions:
	
	  - Set up a time-triggered action to automatically approve the new value if the sales manager doesn't approve or deny it within one calendar day.
	
	  - Gotcha: Time-dependent actions require careful setup and testing to ensure they trigger at the right time.

6. Adjust security and permissions:

- Ensure that the proper profiles and permission sets have access to initiate and approve the change.

- Gotcha: Incorrect permissions can either expose the process to unauthorized users or prevent authorized users from accessing it.

7. Configure Approval Page Layouts (if needed):

- Customize the layout of the approval request so that it includes relevant information.

8. Test the Approval Process:

- Test the process in a sandbox or developer environment to ensure it works as expected. Test with different profiles, time zones, and so on.

- Gotcha: Thorough testing is essential to uncover any hidden issues or misunderstandings in the requirements.

9. Activate the Approval Process:

- Once thoroughly tested, activate the approval process.

10. Document and Train Users:

- Provide documentation and training to end users to ensure they understand the new process.

- Gotcha: Lack of training may result in confusion and reduced adoption.

Note: It would be beneficial to have detailed information regarding the territories, the relationship between the contacts and the sales managers, and any specific organizational policies that need to be adhered to. Understanding these factors upfront can help in crafting a solution that aligns with the actual business needs, minimizing iterations, and rework.

By meticulously addressing these areas, the implementation should align with the goal and acceptance criteria stated in the user story, allowing for a seamless testing and acceptance process.

For an experienced developer, these step-by-step directions are all that is needed to create the approval process. For a less experienced developer, these step-by-step directions provide guidance and, when a particular step isn't clear, the developer can ask ChatGPT to expand that step.

Summary

ChatGPT serves as a valuable resource for developers by "leveling up" the developers with less experience and helping to narrow the knowledge gap between those with differing experience levels. Offering quick and relevant insights enables less experienced developers to access information that might typically require several more years to learn. This access to information can aid in problem-solving and allows for more informed decision-making within development teams. While not a replacement for hands-on experience, ChatGPT can assist by providing guidance and support to those looking to enhance their skills.

The next chapter explores Salesforce Flows and how ChatGPT can streamline the creation, optimization, and troubleshooting of these automated processes. It begins by introducing Salesforce Flows and ChatGPT's role in enhancing flow design and management. The chapter then delves into the practical aspects of designing efficient and intuitive Flows with ChatGPT's assistance, covering everything from basic setups to complex configurations. Significant focus will be put on troubleshooting, where ChatGPT's capabilities in diagnosing and resolving Flow errors will be highlighted. This approach will equip you with the knowledge and skills to leverage ChatGPT effectively in enhancing Salesforce Flow efficiency and user experience.

3
Using ChatGPT for Salesforce Flows

With Flows, admins and developers can create reusable processes, handle more complex scenarios, and even interact with external systems. Flows are a powerful tool, creating powerful avenues for automation within Salesforce.

Flows can be integrated with Apex. This integration enables the optimization of both tools, allowing Flows to be used for most automation needs while reserving Apex for scenarios requiring extensive customization.

Salesforce Flows introduced a significant innovation in business automation by offering a visual approach to defining logical operations, thus reducing the need for manual coding. However, this seemingly simple graphical interface translates into an extremely complex XML structure when represented in text form. The complexity is a result of the relationships, attributes, and elements that are easily understood within a graphical user interface but become challenging to decode in XML.

This complexity presents certain challenges when using automation tools such as ChatGPT to generate Flow XML, often resulting in inaccurate or unworkable code. The intricate relationships and conditions embedded within a Flow are difficult to articulate in a text-based format, making them challenging for an AI model such as ChatGPT to comprehend. The task becomes further complicated when considering the validation and debugging process of the generated XML, rendering it impractical to rely solely on ChatGPT for this task.

Despite these challenges, ChatGPT remains an invaluable resource for Flow design and development guidance. It can offer insights into best practices, troubleshoot common errors, and provide an understanding of optimal design patterns for specific use cases.

This chapter provides a comprehensive look at how ChatGPT can facilitate the creation, optimization, and troubleshooting of Salesforce Flows. It addresses the inherent complexity of setting up these automated processes and demonstrates how AI can simplify the creation of effective workflows tailored to business needs. You will learn how to integrate ChatGPT's capabilities from the initial stages of understanding and designing Flows to handling complex challenges that arise. This chapter will equip you with the knowledge to leverage ChatGPT for efficient flow design, problem-solving, and advanced flow configurations, with a practical case study reinforcing these concepts. By the end of this chapter, you should be able to harness ChatGPT to streamline Salesforce Flows, improving both workflow efficiency and overall user experience.

Unlocking the potential of Salesforce Flows and the role of ChatGPT

An example of a basic Salesforce Flow is the creation of a follow-up task when certain conditions are met on a record, such as a change in the status of an Opportunity.

When a sales organization wants to ensure that every time an Opportunity reaches the "Negotiation" stage, the sales rep's finance team representative is added to the opportunity team, and a notification email is sent to the sales rep and the finance team representative.

Let's look at the two user stories for this requirement.

User story 1 – *when an Opportunity stage is set to Negotiation, add the sales reps' finance team representative to the opportunity team*:

- **Goal**: As a sales rep, I want my opportunities at the Negotiation stage to have my finance team representative on the opportunity team

- **Acceptance criteria**:

  - **Given** there is an opportunity in Salesforce

  - **When** the opportunity stage is changed to "Negotiation"

  - **Then** the finance team member supporting the opportunity owner is added to the opportunity team

User story 2 – *when an Opportunity stage is set to Negotiation, send an email to the sales rep and the finance team representative letting them know about the opportunity stage being set to Negotiation*:

- **Goal**: As a sales rep, I want my finance team representative to know when an opportunity stage has been set to "Negotiation" and the finance team representative has been added to the opportunity team

- **Acceptance criteria**:

    - **Given** there is an opportunity in Salesforce

    - **When** the opportunity stage is changed to "Negotiation"

    - **Then** an email is sent to the opportunity owner and finance team member supporting the opportunity owner, letting them know that the opportunity stage has been set to "Negotiation" and the finance team representative has been added to the opportunity team

When providing a prompt to ChatGPT that includes a user story, the first part of the prompt is a request that frames the overall ask. For user stories that can be addressed with a flow, a good starting point for the request is as follows:

"Please provide detailed directions on how to create a Salesforce Flow to address the following user story. Please note any gaps in the user story that leave an opportunity for misunderstandings, as well as any parts of the flow that may be tricky to create. Include error handling in the Flow, and write a description of what the Flow does for the project documentation."

Let's combine the request and the first user story:

Prompt	Please provide detailed directions on how to create a Salesforce Flow to address the following user story. Please note any gaps in the user story that leave an opportunity for misunderstandings, as well as any parts of the flow that may be tricky to create. Include error handling in the Flow, and write a description of what the Flow does for the project documentation.
	User story 1 – when an Opportunity stage is set to Negotiation, add the sales reps' finance team representative to the opportunity team:
	<ul><li>**Goal**: As a sales rep, I want my opportunities at the Negotiation stage to have my finance team representative on the opportunity team</li><li>**Acceptance criteria**:<ul><li>**Given** there is an opportunity in Salesforce</li><li>**When** the opportunity stage is changed to "Negotiation"</li><li>**Then** the finance team member supporting the opportunity owner is added to the opportunity team</li></ul></li></ul>

Response	Certainly! Below are the detailed directions to create a Salesforce Flow to address the given user story. This guidance also includes potential gaps, trickier parts, and a description for the project documentation. Creating the Flow 1. Navigate to Flow Builder: In the Setup, search for "Flows" and select the Flow Builder. 2. Create a New Flow: Click on "New Flow" and select "Record-Triggered Flow" since this Flow triggers on a record change. 3. Trigger Configuration: Configure the Flow to start when an Opportunity record is updated. Add a condition to check that the Opportunity stage is changed to "Negotiation." 4. Find the Sales Rep's Finance Team Representative: This can be the tricky part. You may need to query the User object or a custom object that stores the relationship between Sales Reps and their finance team representatives. Be sure to handle potential errors if the finance representative is not found. 5. Create Opportunity Team Member: Create a new Opportunity Team Member record, setting the Opportunity ID to the current Opportunity and the User ID to the found finance team representative. 6. Error handling: Add error-handling elements to the Flow. For example, use a Fault connector to send an email to an administrator if the Flow fails at any point. 7. Save and activate: Once the Flow has been correctly configured, save and activate it. Gaps and potential misunderstandings: • Finance team relationship: The user story assumes a specific relationship between the sales reps and their finance team representatives, but it does not detail how this relationship is represented in Salesforce. This may lead to misunderstandings and requires a clear definition in Salesforce. • Opportunity team configuration: The opportunity team configuration might vary between Salesforce implementations. If specific roles or access levels need to be assigned to the finance team representative, these should be specified in the user story. • Error handling: While basic error handling is included, specific business rules for handling exceptions might need to be detailed in the requirements.

Project documentation description:

This Flow automates the addition of a sales rep's finance team representative to the opportunity team when an Opportunity stage is set to "Negotiation." It queries the relationship between the sales rep and their finance team representative and creates a new opportunity team member record. Error handling is included to notify administrators if the Flow encounters any issues. This automation ensures that the finance team is consistently involved at the correct stage of the Opportunity process, enhancing collaboration and ensuring alignment with financial considerations.

By adhering to these detailed directions and carefully considering the potential gaps and complexities, this Flow should effectively meet the user story's requirements, streamlining a critical aspect of the Opportunity management process.

Highlighting the gaps and potential misunderstandings in the response is an essential aspect of creating robust and error-free Salesforce Flows. By identifying these areas of ambiguity or complexity upfront, developers and stakeholders can address them proactively rather than reacting to issues after they've occurred. This anticipatory approach contributes to a more streamlined development process and a solution that's more closely aligned with business needs.

For example, in the context of the described user story, the relationship between sales reps and their finance team representatives was noted as a potential gap. Addressing this might involve creating a clear mapping or relationship within Salesforce that defines this association. Clear documentation, collaboration with stakeholders, and perhaps additional user stories or requirements could be used to fill this gap. Similarly, providing more details on the opportunity team configuration could prevent misunderstandings about roles and access levels. A documented approach to Flow error handling for the project would result in consistency across the flows.

Addressing these issues early in the development process can save time, reduce the risk of errors, and foster a more collaborative and clear understanding of the project's goals and constraints. It establishes a solid foundation for the development, testing, and acceptance of the solution.

A powerful feature of ChatGPT is its ability to drill down into the responses. *Step 5* in the response is as follows.

5. Create an opportunity team member: *Create a new opportunity team member record, setting the Opportunity ID to the current Opportunity and the User ID to the found finance team representative.*

Let's ask ChatGPT to provide more details about this step:

Prompt	Please provide more detail and step-by-step directions in Step 5. Create an opportunity team member.

Response	Certainly! In the context of the given user story, Step 5, creating an opportunity team member, is a crucial aspect. Here's a detailed step-by-step guide on how to execute this in Salesforce Flow:
	
	1. Find the Create Records element: Within your Flow, look for the Create Records element in the toolbox. Drag and drop this element onto the canvas.
	
	2. Name the element: Give the element a meaningful name, something like Create_Opportunity_Team_Member to ensure it's clear what this step is doing.
	
	3. Set the object: In the Object field, select Opportunity Team Member as the object you want to create.
	
	4. Set the field values: You will need to set the values for the fields that define the opportunity team member record:
	
	   • OpportunityId: Reference the Opportunity ID you are working with. This would typically come from a variable or a record that you queried earlier in the Flow.
	
	   • UserId: Reference the ID of the finance team member that you identified in earlier steps. This could involve querying based on the Opportunity owner's finance team representative relationship.
	
	   • TeamRoleId: If you have predefined roles within the opportunity team, you might set this value based on the role the finance team representative should have.
	
	5. Configure additional options: Depending on your specific requirements, you might set additional options, such as how to handle null values or what to do if the creation fails.
	
	6. Error handling: Consider how you will handle any errors that might occur during this creation. You can use the Fault connectors in Flow to direct the logic to an error-handling routine if something goes wrong.
	
	7. Connect to previous and next steps: Make sure the Create Records element is properly connected to the preceding and following elements in your Flow. The previous step should guide into this creation step, and upon successful creation, the logic should move to the next appropriate step in the Flow.

8. Debug and test: Use the debug options within Flow Builder to test this specific step and ensure that the opportunity team member is being created as expected. Check for any issues and correct them as needed.

9. Document: Consider adding comments or annotations within Flow Builder to document what this step is doing, why it's necessary, and any specific considerations or gotchas. Good documentation is essential for maintaining and extending the Flow in the future.

By following these detailed steps, you should be able to create an opportunity team member as part of the process to add the sales reps' finance team representative to the opportunity team when the Opportunity stage is set to Negotiation. This fulfills an essential part of the given user story and aligns the automation with the desired business process.

The ability of ChatGPT to help a developer quickly drill down into specific aspects of a task and elucidate it in detail showcases its true power as an educational and developmental tool. For a seasoned developer, a top-level overview might be sufficient to proceed with the implementation, but the real strength of ChatGPT becomes evident when catering to developers earlier in their careers. They can take advantage of the detailed guidance to explore the nuances of tasks such as creating a Salesforce Flow, which might be unfamiliar or complex.

By providing tiered insights, ChatGPT ensures that no matter the experience level, each developer is quickly equipped with the right information at the right time. Experienced professionals can quickly verify their understanding and approach, while newer developers can delve deeper, accessing mentor-like guidance that builds their confidence and expertise. This bridges the knowledge gap, enhancing productivity and accelerating the learning curve.

Moreover, ChatGPT's approach fosters an inclusive environment where questions are not just answered, but underlying concepts are clarified. The ability to engage with the system for specific guidance imitates a real-world mentorship experience. By facilitating this detailed exploration and understanding, ChatGPT stands as a robust ally for developers at all stages of their careers, harmonizing the workflow and ensuring that the complex becomes accessible.

For developers who want to learn as they work on the user stories assigned to them, ChatGPT can also recommend reading and learning materials:

Prompt	Please recommend reading and learning materials that can be used to understand and do the activities listed in these two responses.
Response	Certainly! To understand and execute the activities listed in the previous discussion, focusing on Salesforce Flows, Apex coding, and the distinctions between them, you might find the following reading and learning materials valuable: 1. Trailhead by Salesforce: For Flows: Modules such as Flow Builder Basics and Automate Your Business Processes with Lightning Flow will offer hands-on experience For Apex: Trailhead offers the Apex Basics & Database and Apex Triggers modules, which provide a solid foundation 2. Salesforce developer documentation: For Flows: Explore Flow Developer Guide, which covers everything from basic concepts to advanced techniques For Apex: Apex Developer Guide is an exhaustive resource for Apex coding 3. Salesforce blogs and community forums: Engage with community blogs and forums that provide real-world examples, tips, and insights from Salesforce professionals 4. Pluralsight courses: Getting Started with Programming in Apex for those looking to learn Apex Salesforce Administration: The Big Picture covers various automation tools, including Flows 5. Udemy courses: Complete Salesforce Lightning Experience User Training covers different Salesforce automation tools The Complete Salesforce Classic Administrator Guide will provide an excellent understanding of Apex 6. Books: Development with the Force.com Platform: Building Business Applications in the Cloud, by Jason Ouellette, provides a comprehensive guide for Apex Mastering Salesforce DevOps, by Andrew Davis, offers insights into best practices, including when to use Flows or Apex

7. YouTube channels:

Salesforce's official channel and other community channels often provide video tutorials and webinars on Flows, Apex, and more

8. Meetups and Salesforce user groups:

Engaging with local Salesforce developer groups or attending Dreamforce and other Salesforce events will provide networking and learning opportunities

9. Online blogs and websites:

Follow Salesforce MVPs and recognized community contributors who regularly share insights and tutorials on these topics

10. Formal Salesforce certifications:

Consider pursuing Salesforce certifications such as Platform Developer I and II or the Advanced Administrator certification to have structured learning paths

These resources range from beginner to advanced levels, providing ample opportunity for professionals at different stages of their Salesforce journey to learn and grow. By combining official documentation with community insights and hands-on practice, developers can obtain a well-rounded understanding of the Salesforce ecosystem.

The ability of ChatGPT to recommend reading and learning materials tailored to specific challenges within Salesforce, as well as for overall learning, is indeed a remarkable advancement in knowledge dissemination. As a platform that unites deep technical understanding with real-world applicability, it aligns perfectly with the practical and strategic needs of professionals in the Salesforce ecosystem.

For those wrestling with complex problems in Apex coding, Lightning Web Components, or integration scenarios, ChatGPT's recommendations can act as laser-focused guides, zeroing in on precisely the documentation, tutorials, or community threads that elucidate the problem. This translates into an acceleration in problem-solving, sparing the time and effort typically expended in sifting through irrelevant material. At a broader level, for those seeking to deepen their understanding of Salesforce's vast suite of products and capabilities, ChatGPT's informed guidance can shape learning paths that are coherent, progressive, and aligned with industry trends and needs. This synergy between targeted problem-solving support and overarching learning facilitation is emblematic of a mature, responsive, and insightful mentorship model. It's not just about answering questions; it's about building competence and confidence, underpinning not just immediate success but long-term growth within the dynamic and ever-evolving Salesforce landscape.

While there are two user stories, each describing a different facet of the requirement, they can be implemented in one flow. This is another area where the power of ChatGPT comes into play.

Let's update the standard request language so that it addresses two user stories with one set of directions.

Prompt	Please provide detailed directions on how to create a Salesforce Flow to address the following two user stories. Please note any gaps in each of the user stories as well as any gaps across the two user stories that leave an opportunity for misunderstandings, as well as any parts of the recommended flow that may be tricky to create. Include error handling in the Flow, and write a description of what the Flow does for the project documentation.
	User story 1 – when an Opportunity stage is set to Negotiation, add the sales reps' finance team representative to the opportunity team:
	• **Goal**: As a sales rep, I want my opportunities at the Negotiation stage to have my finance team representative on the opportunity team
	• **Acceptance criteria**:
	▪ **Given** there is an opportunity in Salesforce
	▪ **When** the opportunity stage is changed to Negotiation
	▪ **Then** the finance team member supporting the opportunity owner is added to the opportunity team
	User story 2: When an Opportunity stage is set to N egotiation, send an email to the sales rep and the finance team representative letting them know about the opportunity stage being set to negotiation:
	• **Goal**: As a sales rep, I want my finance team representative to know when an opportunity stage has been set to Negotiation and the finance team representative has been added to the opportunity team
	• **Acceptance criteria**:
	▪ **Given** there is an opportunity in Salesforce
	▪ **When** the opportunity stage is changed to Negotiation
	▪ **Then** an email is sent to the opportunity owner and finance team member supporting the opportunity owner letting them know that the opportunity stage has been set to "negotiation" and the finance team representative has been added to the opportunity team

Response	Certainly! Let's start by identifying any gaps or ambiguities in the user stories, and then proceed to create a detailed guide to address both of them through a Salesforce Flow. User stories analysis User story 1 Gaps and ambiguities: How is the relationship between the sales rep and the finance team representative defined? What specific role should the finance team member have within the opportunity team? Are there any permissions or restrictions to consider? User story 2 Gaps and ambiguities: What specific information should be included in the email? Is there a specific email template to be used? Gaps across both user stories: Both stories are interdependent but are not linked in their description The process of identifying the appropriate finance team member is not detailed Creating a Salesforce Flow: 1. Define the trigger: Set the flow to trigger when an Opportunity stage changes Check if the new stage is set to Negotiation 2. Add a finance team member to the opportunity team (user story 1): Get the opportunity owner: Retrieve the sales rep (opportunity owner). Find finance team member: Query the finance team member related to the sales rep (note: the mapping must be pre-defined or determined via a formula or lookup) Create an opportunity team member record: Add the finance team member to the opportunity team, assigning the appropriate role. Handle errors with a fault path, logging an error message if this process fails.

3. Send an email notification (user story 2):

Prepare email content: Utilize an existing template or dynamically create the email content, including details about the opportunity and the change to the Negotiation stage.

Send Email: Use the Send Email action to dispatch the email to the sales rep and the finance team member. Include error handling for potential failures in sending the email.

Project documentation description:

This Flow facilitates seamless collaboration between sales reps and their corresponding finance team representatives within the Salesforce Opportunities process. Triggered by the Opportunity stage being set to Negotiation, the Flow automatically adds the related finance team member to the opportunity team and sends an email notification to both parties. This ensures that both the sales rep and the finance team representative are aligned and informed during crucial negotiation stages, thereby fostering synergy and efficiency in deal closing.

Tricky parts and recommendations:

Ensuring that the relationship between the sales rep and the finance team member is accurately mapped and retrieved can be challenging. This might require a custom object or a specific field on the User object.

Care must be taken in defining the correct permissions for modifying the opportunity team. The running user of the Flow must have the appropriate access rights.

Robust error handling is included in the Flow to manage any unexpected issues, such as failure to find the finance team member or in sending the email.

By addressing the potential ambiguities and gaps in the user stories, this Flow creates a unified process that aligns with the functional and communication needs specified, enhancing the collaboration within the Salesforce Opportunity management life cycle.

It is crucial to approach ChatGPT not with a presumption of limitations but rather as an engaged partner akin to an experienced developer. This mindset fosters a dynamic collaboration where the potential to support flow development and other complex Salesforce tasks can be fully explored. By initiating the dialogue with a willingness to probe, experiment, and iterate through various prompts, you unlock the opportunity to leverage ChatGPT's extensive knowledge and analytical prowess. Only after a comprehensive and nuanced exploration, if gaps are identified, should you transition to addressing the work independently of ChatGPT. This approach not only maximizes the utility of this AI tool but also fosters a culture of curiosity, innovation, and collaboration that resonates with the broader vision of harnessing generative AI to empower human creativity and problem-solving within the Salesforce ecosystem.

Troubleshooting Flow errors with ChatGPT assistance

Developers often encounter Flow development and support challenges that require effective troubleshooting. In this section, you will gain a brief insight into typical errors that may arise within Salesforce Flows, including challenges related to governor limits, unhandled faults, and permission issues. Guidance on how to interact with ChatGPT will be provided, including instructions on how to formulate questions and interpret its responses to address specific Flow problems. Step-by-step guides and practical examples will demonstrate how you can utilize ChatGPT in solving real-world issues encountered in Flow creation and maintenance. The aim is to equip Salesforce developers with an additional tool that combines human expertise with the capabilities of ChatGPT to enhance problem-solving within the Salesforce environment.

Governor limits

Salesforce enforces various governor limits to control the resources that are consumed by Flows, such as SOQL query limits, DML operation limits, and more. A sample error message is *Too many SOQL queries: 101.*

Prompt	I am working on a Salesforce Flow and I am getting the following error. What could be causing this error and what steps should I follow to troubleshoot this error? Please recommend reading and learning materials I can review to learn more about this error and how to remediate it.
	Too many SOQL queries: 101.
Response	Ah, the infamous "Too many SOQL queries: 101" error! This is a common governor limit error in Salesforce, particularly when dealing with Flows, triggers, or classes. Let's break down the cause and steps to troubleshoot this error.
	Cause:
	Salesforce imposes a governor limit of 100 SOQL queries in a single transaction. When this limit is exceeded, the error "Too many SOQL queries: 101" is thrown. It might occur when loops or recursive calls are used excessively, causing a large number of SOQL queries within a single execution context.
	Steps to troubleshoot:
	1. Identify the source: Check the Flow or code that's causing the error. Look for any loops or repetitive SOQL queries that might be running within the transaction.
	2. Optimize the queries: Rather than querying inside a loop, aim to bulkify your queries. Gather all the necessary IDs or conditions first, then execute the SOQL query outside the loop.

3. Utilize debug logs: Set up debug logs to trace the execution and pinpoint exactly where the limit is being reached.

4. Test in different contexts: Run tests in different user contexts or with different datasets to understand how the queries behave under various scenarios.

5. Consider using batch or asynchronous processing: If the logic inherently requires a large number of queries, consider implementing it in batch classes or using asynchronous methods that separate the transactions.

Recommended reading and learning materials:

Salesforce Developer Documentation on Governor Limits: This is an authoritative source to understand all the governor limits, including SOQL query limits

Trailhead Module on Bulk Apex Trigger: This module covers bulkifying code, a technique that will help you avoid such SOQL limit errors

Apex Developer Guide's Section on SOQL Query Optimization: This provides detailed insights into writing efficient SOQL queries

With the error you're facing, a blend of understanding the Salesforce platform's constraints and a thoughtful examination of your Flow or code will lead you to an optimized solution. By aligning your development approach with these best practices, you'll foster robust and efficient implementations within the Salesforce ecosystem.

This response is somewhat generic as it addresses the commonly encountered "Too many SOQL queries: 101" error within the context of Salesforce without specific details about the specific Flow functionality. To receive a more focused and tailored solution, it would be helpful to include information about what the Flow is designed to achieve. For instance, in this scenario, the specific Flow's purpose could be to update the custom **Account Address** field for all of the account's contacts when the account address is changed. The error seems to manifest when an account has more than 100 contacts. Such detailed insight into the Flow's functionality and the specific conditions under which the error occurs would enable a more targeted analysis and recommendation, aligning closely with the unique requirements and constraints of the given Salesforce implementation.

> **Note**
> Getting to the bottom of the error and the best fix may require several refining prompts. ChatGPT never gets tired of the same question being asked in different ways, and one of the ways of asking may be what provides the response that fills the gaps in your knowledge.

An unhandled fault has occurred in this flow

This generic error might appear if a record is not found or a null value is processed inappropriately. Let's take a look:

Prompt	I am working on a Salesforce Flow. The Flow is triggered when an account address is updated and the flow updates a custom address field in every contact record associated with the account. I am getting the following error. What could be causing this error and what steps should I follow to troubleshoot this error? Please recommend reading and learning materials I can review to learn more about this error and how to remediate the error.
	"An unhandled fault has occurred in this flow."
Response	The error message "An unhandled fault has occurred in this flow" in Salesforce is quite generic and can be caused by a variety of underlying issues. In your specific context, where the Flow is triggered by an update to an account's address and consequently updates a custom address field for all associated contacts, let's break down the potential causes and steps to troubleshoot this error.
	Potential causes:
	1. Field accessibility: If the custom address field is not accessible to the user executing the Flow, an unhandled fault might occur.
	2. Null or invalid values: If the account or contacts have null or unexpected values in the address fields that the Flow is expecting, this could lead to an error.
	3. Bulk data issues: If there are a large number of contacts related to the account, this might be causing the Flow to exceed governor limits.
	Steps to troubleshoot:
	1. Inspect the fault email: Salesforce sends an email containing details of the fault. This is your starting point to understand what went wrong.
	2. Check field accessibility: Ensure that the user running the Flow has the necessary permissions to access and modify the custom address field.
	3. Review data validity: Verify that the data in the account and contact fields that are being processed by the Flow meet the expected format and are not null.
	4. Utilize debug logs: Activate debug logs for the user running the Flow to trace the execution and pinpoint the exact place where the fault is occurring.
	5. Test with different scenarios: Run tests with different datasets and user profiles to reproduce and understand the error under various conditions.

6. Consider implementing error handling: Building in fault handling within the Flow can make it more robust and provide more specific error messages in the future.

Recommended reading and learning materials:

Salesforce Developer Documentation on Debugging Flows: Detailed guidelines on using debug logs for Flows

Trailhead Module on Flow Error Handling: A hands-on guide to adding fault handling to your Flows

Salesforce Developer Community: Engage with the community to discuss specific issues and best practices related to Flow development

By following these steps and utilizing the recommended resources, you should be able to diagnose and remediate the error in your Flow. Since the error message is generic, a systematic approach that includes reviewing specific error details, checking permissions, validating data, and leveraging debug logs will be essential to pinpointing and resolving the underlying issue.

As with the previous defect, the initial response to the error "An unhandled fault has occurred in this flow" is necessarily a bit generic, given the broad nature of the problem statement. ChatGPT serves as a valuable partner in the troubleshooting process, and it can assist you in narrowing down the root cause through an iterative dialogue. As you begin to identify and discard potential causes, you can engage with ChatGPT using a series of refining prompts. Each subsequent prompt should contain more specific details about the problem or the findings from previous investigation steps. In this manner, ChatGPT can progressively hone in on the precise issue, offering more focused guidance and solutions. It's a collaborative process that leverages the power of generative AI to support your Salesforce development efforts, aligning closely with real-world troubleshooting methodologies.

Permission and sharing issues

Flows run in different contexts (for example, user mode or system mode). This can lead to unexpected permission issues, especially when dealing with object-level or field-level security. Here's the error message:

"*INSUFFICIENT_ACCESS_ON_CROSS_REFERENCE_ENTITY, insufficient access rights on cross-reference id.*"

Let's take a closer look:

Prompt	I am working on a Salesforce Flow. The Flow is triggered when an account address is updated and the flow updates a custom address field in every contact record associated with the account. The permissions in the Salesforce org are very tightly managed. I am getting the following error. What could be causing this error and what steps should I follow to troubleshoot this error? Please recommend reading and learning materials I can review to learn more about this error and how to remediate the error. INSUFFICIENT_ACCESS_ON_CROSS_REFERENCE_ENTITY, insufficient access rights on cross-reference id.
Response	The error message "INSUFFICIENT_ACCESS_ON_CROSS_REFERENCE_ENTITY, insufficient access rights on cross-reference id" in Salesforce is indicative of a permissions issue. Since you've already pointed out that permissions in the Salesforce org are very tightly managed, it aligns with the nature of this error. Here's a breakdown of potential causes and troubleshooting steps for this specific scenario. Potential causes: 1. Field-level security (FLS): The running user may not have the required permissions to access or modify the custom address field. 2. Object-level security: The user may lack the necessary permissions on the Account or Contact objects themselves. 3. Sharing rules: If the Account or Contact objects are owned by another user, sharing rules might prevent the running user from accessing or modifying those records. Steps to troubleshoot: 1. Verify FLS: Check that the user profile or permission set includes access to the custom address field that you're attempting to update. 2. Inspect object-level permissions: Ensure that the user has the proper read and write permissions on both the Account and Contact objects. 3. Review sharing settings: Examine the sharing rules for the Account and Contact objects, and ensure that the running user has the required access level. 4. Utilize debug logs: Enable debug logs for the user executing the Flow to get detailed insights into what's happening at runtime. 5. Consider manual testing: Try manually updating the field through the UI logged in as the user in question to reproduce the error and gain insights.

> Recommended reading and learning materials:
>
> Salesforce Documentation on Understanding Sharing: An in-depth guide to understanding and configuring sharing settings
>
> Trailhead Module on Data Security: This module will give you a hands-on understanding of object-level and field-level security
>
> Salesforce Developer Community Forums: You may find discussions related to similar issues and resolutions
>
> The error you've encountered is tied to the permissions model within your Salesforce organization. By methodically examining field-level, object-level, and sharing settings, and by using tools such as debug logs, you can identify and correct the specific access issue at play. Leveraging these techniques and resources will enable you to resolve the error and align the Flow's behavior with your organization's tightly managed permissions architecture.

The ChatGPT response to the specific error message and Flow was noticeably more focused, likely owing to the detailed information provided about the Salesforce org and the particular nature of the Flow. This illustrates a fundamental principle in engaging with generative AI such as ChatGPT: the more context and specificity you provide, the more precise and tailored the guidance becomes. Just as ChatGPT can be a vital partner in creating Flows, benefiting from a series of refining prompts to nail down the details of what needs to be crafted, it also serves as an invaluable ally in troubleshooting Flows. Through a collaborative and iterative approach, utilizing successive prompts to build upon previous insights, you and ChatGPT can work together to methodically analyze and resolve issues. This synergy can significantly enhance the efficiency and effectiveness of both the development and troubleshooting processes within the Salesforce ecosystem.

Advanced flow techniques with ChatGPT

Always consider using stretch goals when engaging with ChatGPT, especially in the context of Salesforce Flow development. Think of ChatGPT as a senior developer within your team; if a question or challenge is something you would reasonably present to an experienced colleague, don't hesitate to bring it to ChatGPT as well. By pushing the boundaries and seeking insights into intricate problems, you may discover solutions and perspectives that you hadn't considered. Embracing this approach not only leverages the advanced capabilities of ChatGPT but also fosters a dynamic and collaborative environment that enhances problem-solving and innovation. Utilizing ChatGPT as an extension of your development team can amplify your resources and contribute to achieving both the immediate objectives and the broader goals of your projects.

Let's look at three user stories related to Opportunity stages.

User story 1 – *when an Opportunity stage is "Decision Makers," use part of the Opportunity page to display the contacts associated with the account and require the sales rep to select at least one decision maker for the opportunity*:

- **Goal**: As a sales rep, I want to select at least one decision-maker for the opportunity
- **Acceptance criteria**:
  - **Given** there is an opportunity in Salesforce
  - **When** the opportunity stage is set to "Decision Makers"
  - **Then** a list of contacts is displayed and the user can select a decision-maker

User story 2 – *when an Opportunity stage is "Perception Analysis," use part of the Opportunity page to display a list of company executives and enable the sales rep to select one to call the opportunity decision-maker*:

- **Goal**: As a sales rep, I want to select a company executive to call the opportunity decision maker(s) to help close the deal
- **Acceptance criteria**:
  - **Given** there is an opportunity in Salesforce
  - **When** the opportunity stage is set to "Perception Analysis"
  - **Then** a list of company executives is displayed and the user can select a company executive

User story 3 – *when an Opportunity stage is "Negotiation," use part of the Opportunity page to display the discounts the user can offer the decision maker. For each discount, include the amount by which the sales rep's commission is reduced*:

- **Goal**: As a sales rep, I want to see a list of the discounts that can be offered to the decision-maker, along with the reductions in my commission
- **Acceptance criteria**:
  - **Given** there is an opportunity in Salesforce
  - **When** the opportunity stage is set to "Negotiation"
  - **Then** a list of discounts and commission reductions is listed

Let's write a request to have ChatGPT recommend a Flow-based approach to these user stories:

Please recommend a Flow-based approach to address the following user stories:

Prompt	Please recommend a Flow-based approach to address the following user stories.
	User story 1 – when an Opportunity stage is "Decision Makers," use part of the Opportunity page to display the contacts associated with the account and require the sales rep to select at least one decision maker for the opportunity:
	• Goal: As a sales rep, I want to select at least one decision-maker for the opportunity
	• Acceptance criteria:
	▪ Given there is an opportunity in Salesforce
	▪ When the opportunity stage is set to "Negotiation"
	▪ Then a list of contacts is displayed and the user can select a decision-maker
	User story 2 – when an Opportunity stage is "Perception Analysis," use part of the Opportunity page to display a list of company executives and enable the sales rep to select one to call the opportunity decision-maker:
	• Goal: As a sales rep, I want to select a company executive to call the opportunity decision maker(s) to help close the deal
	• Acceptance criteria:
	▪ Given there is an opportunity in Salesforce
	▪ When the opportunity stage is set to "Perception Analysis"
	▪ Then a list of company executives is displayed and the user can select a company executive
	User story 3 – when an Opportunity stage is "Negotiation," use part of the Opportunity page to display the discounts the user can offer the decision maker. For each discount, include the amount by which the sales rep's commission is reduced:
	• Goal: As a sales rep, I want to see a list of the discounts that can be offered to the decision-maker, along with the reductions in my commission
	• Acceptance criteria:
	▪ Given there is an opportunity in Salesforce
	▪ When the opportunity stage is set to "Negotiation"
	▪ Then a list of discounts and commission reductions is listed

Response	Dynamic Actions in Salesforce offer a way to display or hide actions based on record data and user information. This capability can be highly effective in addressing the scenarios described in the user stories.
	User story 1:
	1. Flow trigger: When the Opportunity stage is set to Decision Makers.
	2. Dynamic Actions: Configure Dynamic Actions to display the associated contacts section based on the Opportunity stage.
	3. Flow action: Implement a screen Flow to present the contacts associated with the Opportunity's account.
	4. User interaction: Enable the sales rep to select at least one decision maker.
	5. Validation: Validate that at least one decision maker is selected.
	6. Update record: Associate the selected decision maker(s) with the Opportunity.
	User story 2:
	1. Flow trigger: When the Opportunity stage is set to Perception Analysis.
	2. Dynamic Actions: Utilize Dynamic Actions to reveal the company executives' section based on the stage.
	3. Flow action: Use a screen Flow to display a list of company executives.
	4. User interaction: Allow one executive to be selected.
	5. Update record: Store the selected executive's information with the Opportunity.
	User story 3:
	1. Flow trigger: When the Opportunity stage is set to Negotiation.
	2. Dynamic Actions: Deploy Dynamic Actions to dynamically show the discounts and commission reductions section based on the stage.
	3. Flow action: Implement a screen Flow to display discounts and corresponding commission reductions.
	4. User interaction: Enable viewing and selection of a discount offer.
	5. Calculations and logic: Apply calculation logic to compute commission reductions.
	6. Update record: Update the Opportunity and commission records with the selected discount.

	Conclusion: Dynamic Actions add a layer of interactivity and responsiveness to the solution, allowing for contextual display and control based on the Opportunity stage. This approach fosters an intuitive experience for the sales rep and ensures that the right components are displayed at the appropriate stages of the Opportunity. Leveraging Dynamic Actions within screen Flows aligns with the goals and acceptance criteria of the user stories while taking advantage of Salesforce's modern and flexible flow-building capabilities. It encourages alignment with best practices and opens up possibilities for a richer and more tailored user experience.

ChatGPT, with its vast repository of knowledge and perfect recall, serves as a powerful ally in devising complex designs and exploring all available Salesforce features. Its understanding stems from the extensive range of materials that are used for training, enabling it to draw upon a myriad of concepts, technologies, and best practices within the Salesforce ecosystem. However, it's essential to recognize that at the time of writing, ChatGPT's knowledge base is only up to date until Spring 2023. Therefore, certain recent innovations or changes might not be within its current scope. However, the promise of ChatGPT's continued evolution holds the potential for it to become even more relevant and insightful. It's conceivable that updates to ChatGPT will enable it to be more in sync with the very latest Salesforce advancements, offering even more precise and cutting-edge assistance.

Case study

The case study from *Appendix A* has several user stories that can be addressed with Flows.

Let's use the base flow request:

"Please provide detailed directions on how to create a Salesforce Flow to address the following user story. Please note any gaps in the user story that leave an opportunity for misunderstandings, as well as any parts of the flow that may be tricky to create. Include error handling in the Flow, and write a description of what the Flow does for the project documentation."

We'll also use epic 3 and user story 2 from the case study to investigate ChatGPT's capabilities.

Epic 3 – *email functionality with Snowflake data*

User Story 2 – *create a Flow-based wizard to select a contact*:

- **Goal**: As a sales rep, I want a wizard to guide me through the process of selecting a contact to send the email to so that I can choose the correct recipient.

- **Acceptance criteria**:

  - **Given** a sales rep clicks the **Email** button in the **Snowflake** section

  - **When** the wizard starts

  - **Then** the sales rep should be prompted to select a contact associated with the account

Let's take a look:

Prompt	Please provide detailed directions on how to create a Salesforce flow to address the following user story. Please note any gaps in the user story that leave an opportunity for misunderstandings, as well as any parts of the flow that may be tricky to create. Include error handling in the flow, and write a description of what the Flow does for the project documentation.
	Epic 3 – email functionality with Snowflake data
	User story 2 – create a Flow-based wizard to select a contact:
	• **Goal**: As a sales rep, I want a wizard to guide me through the process of selecting a contact to send the email to so that I can choose the correct recipient
	• **Acceptance criteria**:
	▪ **Given** a sales rep clicks the Email button in the Snowflake section
	▪ **When** the wizard starts
	▪ **Then** the sales rep should be prompted to select a contact associated with the account
Response	Certainly! Let's address the user story and design a Flow-based wizard to select a contact for emailing in Salesforce. We'll also examine any potential gaps and challenges in the process.
	Flow design for email functionality with Snowflake data:
	1. Start the Flow:
	Trigger: The Flow will be triggered by clicking an Email button within the Snowflake section
	Action: A wizard screen will appear, prompting the sales rep to proceed with selecting a contact associated with the account
	2. Screen Element for Contact Selection:
	Create a screen element with a dropdown that lists contacts associated with the account
	This list should be dynamically generated based on the account related to the Snowflake section

3. Error handling and validation:

Include logic to check if there are any contacts associated with the account

If no contacts are found, an error message should be displayed to the sales rep

4. Finish the Flow:

Once the contact has been selected, the wizard should end, and the selection should be available for further processing, such as sending an email

Gaps and challenges:

Ambiguity in the Snowflake section: The story does not clearly define what the "Snowflake section" refers to. Understanding its context is essential to properly link the Email button to the flow.

Selection process clarity: The story does not provide details on what happens after contact selection, which might require collaboration with other stakeholders for complete understanding.

Account association: Ensuring that the listed contacts are relevant to the specific account can be tricky and needs careful handling.

Project documentation

Flow name: Contact selection wizard for Snowflake email functionality.

Description: This Salesforce flow enables sales reps to select a contact associated with an account within the Snowflake section by guiding them through a wizard. Triggered by clicking the Email button, the Flow provides a drop-down list of relevant contacts for selection, includes error handling to deal with scenarios where no contacts are found and ends with the selected contact being available for email communication.

Associated epic: Epic 3 – email functionality with Snowflake data.

User story: User story 2 – create a Flow-based wizard to select a contact.

Goal: To facilitate the correct selection of an email recipient by guiding the sales rep through a structured process.

By carefully considering these steps, challenges, and potential gaps, this Flow aims to provide an intuitive and error-free way for sales reps to select the appropriate contact for their email communications within the Snowflake section. It underscores the importance of alignment with stakeholders to ensure the full context and intended behavior of the Flow are understood and accurately implemented.

ChatGPT identified three gaps and challenges. Let's prompt ChatGPT to refine the answer based on answers to the gaps and challenges:

Prompt	The Snowflake section on the contact page layout is a tab dedicated to displaying information from Snowflake. There are additional user stories that address what happens after this user story. The contacts to be listed are the ones where the account lookup field on the contact is all the same account. With these answers, please update and rewrite the response. Anything unaffected by these answers does not need to change.
Response	Certainly! With the additional context provided, let's refine the Flow design to create a Salesforce Flow-based wizard that aligns with the user story and its requirements.
	Flow design for email functionality with Snowflake data:
	1. Start the Flow:
	Trigger: The flow will be triggered by clicking an Email button within the Snowflake tab on the contact page layout
	Action: A wizard screen will appear, prompting the sales rep to proceed with selecting a contact associated with the account
	2. Screen element for contact selection:
	Create a screen element with a dropdown that lists contacts associated with the same account, where the account lookup field on the contact matches the specific account
	This list should be dynamically generated based on the account related to the Snowflake tab
	3. Error handling and validation:
	Include logic to check if there are any contacts associated with the specific account.
	If no contacts are found, an error message should be displayed to the sales rep
	4. Finish the Flow:
	Once the contact has been selected, the wizard should end, and the selection should be available for further processing, as defined in the subsequent user stories.

Gaps and challenges:

Account association: Ensuring that the listed contacts are relevant to the specific account and filtered by the account lookup field can be tricky and needs careful handling.

Project documentation

Flow name: Contact selection wizard for Snowflake email functionality.

Description: This Salesforce flow enables sales reps to select a contact associated with an account within the Snowflake tab on the contact page layout. Triggered by clicking the Email button, the flow provides a drop-down list of contacts where the account lookup field matches the specific account. It includes error handling to deal with scenarios where no contacts are found and ends with the selected contact being available for further processing as detailed in subsequent user stories.

Associated Epic: Epic 3 – mail functionality with Snowflake data.

User Story: User story 2 – create a Flow-based wizard to select a contact.

Goal: To facilitate the correct selection of an email recipient by guiding the sales rep through a structured process within the Snowflake tab.

By integrating the details about the Snowflake section and the specific Account relationship, this refined flow design aims to provide a seamless experience for Sales Reps in selecting the appropriate contact for email communications. It emphasizes the precision needed in filtering contacts based on the Account lookup field, thus catering to the specific requirements of the given user story.

Even though the "Account Association" concern seems to have been addressed, ChatGPT is still listing it as a concern. This could be a ChatGPT quirk or there could still be a gap. Let's directly ask for an explanation:

Prompt	Under "Gaps and Challenges" is Account Association: ensuring that the listed contacts are relevant to the specific account and filtered by the account lookup field can be tricky and needs careful handling.
	I thought I addressed this by telling you that "The contacts to be listed are the ones where the account lookup field on the contact are all the same account."
	Why is this tricky? What is needed to resolve this "Gap and Challenge"?

Response	My apologies for the confusion. Your direction does indeed clarify the requirement, but let me further explain why I considered this as a potential challenge.
	The "tricky" part I mentioned relates to the design and implementation phase of the flow, specifically in the context of Salesforce. Here's why:
	1. Query complexity: Retrieving contacts based on the specific account lookup field requires a well-constructed SOQL query. While it's a straightforward requirement, the developer must ensure that the query is efficient, especially if there are large numbers of contacts related to an account.
	2. Contextual handling: The Flow must recognize the specific account related to the Snowflake tab from which the Email button is clicked. This requires proper handling of contextual information, ensuring that the correct account ID is used in filtering the contacts.
	3. Dynamic filtering: If the Account field is ever updated, or if the context changes dynamically within the user interface, the flow must respond accordingly, updating the list of contacts. Ensuring that this dynamic filtering works smoothly can require careful testing and validation.
	4. Permissions and security: Ensuring that the user has the appropriate permissions to view and select the contacts related to the specific account must be considered. This involves compliance with the Salesforce org's security model.
	Resolution:
	1. Construct a robust SOQL query: Build a well-designed SOQL query to fetch contacts based on the specific account lookup field.
	2. Contextual Handling: Utilize the context of the button click within the Snowflake tab to identify the relevant account, possibly using Apex if necessary.
	3. Test dynamically: Implement tests to ensure that the dynamic filtering of contacts works as expected as the context changes.
	4. Consider security: Align the flow with the Salesforce org's security protocols to ensure that it adheres to the relevant permissions and profiles.
	In the hands of an experienced Salesforce developer like yourself, these challenges would be well within reach to overcome. I mentioned them to highlight potential areas that might require attention, and they can be seen as aspects to validate rather than insurmountable obstacles.

And since the authors want to get along with our AI overlords, here's the prompt we provided:

Prompt	Thank you for your explanation of how to address this "Gap and Challenge" as well as for your help with this user story!
Response	You're welcome! It's been a pleasure working alongside you on this user story. Your in-depth knowledge and detailed questions have enabled a clear and precise understanding of the requirements. Should you have any more questions or need further assistance, feel free to reach out. Here's to a successful implementation of the Flow, and may it greatly enhance the functionality and user experience within your Salesforce environment. Happy developing!

Summary

You've now traversed the landscape of Salesforce Flows with ChatGPT as your guide. Across this chapter, you've absorbed the essentials of Salesforce Flows, understanding how to employ ChatGPT's insights for more intuitive and sophisticated Flow designs that meet business imperatives. Your newfound skills include the ability to streamline complex automated processes, troubleshoot and resolve flow errors with AI-driven efficiency, and apply advanced techniques to keep your workflows at the cutting edge. The practical case study has helped cement these concepts, illustrating the transformative impact of integrating ChatGPT into the flow configuration and optimization processes. You are now equipped with a robust set of strategies to effectively manage and refine Salesforce Flows, ensuring they align with best practices and contribute to an enhanced operational workflow.

In the next chapter, you will learn how ChatGPT can significantly bolster Salesforce feature design, guiding you through a variety of design approaches to tackle user stories. You'll learn how to harness ChatGPT to generate and meticulously evaluate multiple solution pathways, considering each one's strengths and limitations against a backdrop of Salesforce best practices and practical constraints such as scalability and governor limits. This chapter will not only broaden your design perspective but also sharpen your critical analysis skills, ensuring that you're equipped to make decisions that serve both immediate functional requirements and long-term strategic goals. Expect to gain proficiency in recognizing the nuances of various design approaches, scrutinizing their scalability, and enhancing collaborative decision-making with ChatGPT as a cornerstone of your Salesforce design toolkit.

4

Using ChatGPT for Salesforce Feature Design

Utilizing ChatGPT to evaluate a user story can be a useful aspect of the Salesforce development process. It offers a way to explore different solutions within Salesforce, taking into account the specific requirements of the user story. While it might not provide insights beyond seasoned developers, it serves as a valuable tool to verify or expand on existing ideas.

Scalability is an important factor in Salesforce, and ChatGPT's ability to assess different solutions in this context can help in making informed choices. Although relying solely on ChatGPT might not be sufficient, it can supplement human judgment by providing additional perspectives on how a solution might scale over time. It's important to use this information in conjunction with other assessments to get a well-rounded view of the potential scalability of a given solution.

Using ChatGPT can aid in alignment within a project team, but it shouldn't replace direct communication and collaboration. By listing various design approaches and their pros and cons, it can contribute to the understanding of potential solutions. However, it should be used in conjunction with regular meetings and discussions to ensure that all stakeholders are on the same page.

The tool's ability to quickly enumerate different solutions can be helpful in speeding up the decision-making process. However, it's important to recognize that ChatGPT's insights should be part of a broader evaluation, including human analysis and team discussions. While it can enhance the decision-making process, reliance solely on this tool might not be sufficient for complex and nuanced projects.

ChatGPT can provide access to information and insights that might not be readily available to all team members. While it does not replace the expertise of experienced Salesforce developers, it can be used to supplement knowledge and foster a more informed conversation among team members. It's a tool that should be leveraged in a supportive role rather than as a standalone solution for designing and implementing Salesforce projects.

Prepare to unlock the potential of ChatGPT in articulating a range of design strategies. This chapter guides you through leveraging ChatGPT's vast understanding of detail and critically assessing numerous solutions, aiding you to weigh their benefits and trade-offs effectively. You'll gain insights into integrating broad Salesforce methodologies and best practices into your design choices, focusing on key aspects such as scalability, maintenance, and strategic alignment. By the end of this chapter, you will have sharpened your skills in evaluating design options, understood the importance of scalability, and enhanced your ability to collaboratively make decisions that are robust for the present and flexible for the future of your Salesforce environment.

Exploring multiple solutions with ChatGPT

Getting started with having ChatGPT review a user story and provide solution options is a relatively straightforward process. You'll want to begin by clearly defining the user story (see *Chapter 11*), outlining the specific requirements, constraints, and goals you want to achieve within the Salesforce environment. Then, you can pose a query to ChatGPT that encapsulates the user story, requesting an analysis of potential solutions and their associated pros and cons. It's beneficial to frame the query in a way that encourages a comprehensive response, possibly including questions about scalability, integration with other systems, or adherence to Salesforce best practices. Engaging with ChatGPT in this manner allows you to leverage its vast knowledge base to explore various design approaches and evaluate them according to your project's unique needs. Keep in mind that the insights provided by ChatGPT should ideally be used in conjunction with your expertise to ensure that the chosen solution aligns well with the broader project context and objectives.

Integrating Salesforce and Snowflake

There is a case study in *Appendix A* with epics and user stories. *Epic 1* addresses integrating Salesforce and Snowflake (**Data as a Service**, or **DaaS**). Let's use *Epic 1* to have ChatGPT enumerate the design options for this epic:

Prompt	Please list all of the different ways this epic and these L1 and L2 user stories could be implemented in Salesforce
	
	Epic 1: Integration of Snowflake Data with Salesforce Account Pages:
	
	• Design and implement a point-to-point integration for Salesforce to call Snowflake and retrieve information for a specific Salesforce account when a user views the account page.
	
	User Story 1: Retrieve Snowflake Data for Specific Salesforce Account:
	
	• **Goal:** As a Sales Rep, I want Salesforce to call Snowflake and retrieve information for a specific Salesforce account when I view the account page so that I have real-time insights related to that account.
	
	• **Acceptance Criteria**:
	
	  ▪ **Given** a Sales Rep is viewing a specific Salesforce account page,
	
	  ▪ **When** the page is loaded,
	
	  ▪ **Then** the system should call Snowflake and retrieve the relevant information for that specific account.
	
	User Story 1.1: Establish a Secure Connection with Snowflake:
	
	• **Goal:** As a Sales Rep, I want Salesforce to establish a secure connection with Snowflake when I view an account page, ensuring data integrity and privacy.
	
	• **Acceptance Criteria**:
	
	  ▪ **Given** a Sales Rep is viewing a specific Salesforce account page,
	
	  ▪ **When** the page is loaded,
	
	  ▪ **Then** the system should establish a secure connection with Snowflake.
	
	• Notes: Specific, focused on security, and meets INVEST criteria.

User Story 1.2: Call Out to Request Specific Account Information from Snowflake

- **Goal:** As a Sales Rep, I want Salesforce to make a Call Out to Snowflake to request specific information related to a Salesforce account when I view the account page, providing me with timely insights.

- **Acceptance Criteria**:

  - **Given** a Sales Rep is viewing a specific Salesforce account page and a secure connection to Snowflake is established,

  - **When** the page is loaded,

  - **Then** the system should make a Call Out to Snowflake to retrieve the relevant information for that specific account.

- Notes: Specific, focused on the Call Out mechanism, and meets INVEST criteria.

User Story 1.3: Implement Error Handling for Snowflake Connection and Data Retrieval

- **Goal**: As a Sales Rep, I want any errors that occur during the connection to Snowflake or data retrieval to be handled gracefully so that I am informed about the issue and can continue using the system without disruption.

- **Acceptance Criteria**:

  - **Given** a Sales Rep is attempting to view a specific Salesforce account page,

  - **When** an error occurs in connecting to Snowflake or retrieving data,

  - **Then** the system should log the error, notify the Sales Rep, and allow them to continue using the system without disruption.

- Notes: Specific, focused on error handling, and meets INVEST criteria.

Response	The epic and the associated L1 and L2 user stories you've described can be implemented in Salesforce through various methods, each with distinct considerations. Here's an outline of the different ways this could be accomplished:
	1. Using Salesforce Connect with Custom Adapter:
	Description: Develop a custom adapter for Salesforce Connect to establish a connection with Snowflake. Use External Objects to represent the Snowflake data in Salesforce, allowing real-time querying of data.
	Pros: Simplifies data retrieval, leverages built-in features, and promotes real-time interaction.
	Cons: Requires customization for the specific connection to Snowflake, may have some limitations depending on the complexity of the data.
	2. Utilizing Custom Apex Classes and Callouts:
	Description: Develop custom Apex classes and perform HTTP callouts to interact with Snowflake's APIs. Implement robust error handling and security measures.
	Pros: Highly flexible and customizable, allows for precise control over data retrieval and connection handling.
	Cons: More complex development, potential scalability concerns.
	3. Integration Through Middleware (for example, MuleSoft):
	Description: Utilize middleware such as MuleSoft to facilitate the integration between Salesforce and Snowflake. The middleware acts as a broker, handling communication, data transformation, and error handling.
	Pros: Offers abstraction, simplifies integration complexity, provides robust error handling and transformation capabilities.
	Cons: Introduces additional components, potential latency, costs related to middleware.
	4. Utilizing Salesforce Einstein Analytics with Direct Data Connectors:
	Description: Use Salesforce Einstein Analytics' direct data connectors to Snowflake, pulling data directly into Einstein Analytics and displaying it on the Salesforce account page.
	Pros: Easy to set up, provides robust analytics capabilities.
	Cons: Limited to the capabilities and constraints of Einstein Analytics, potential cost implications.

5. Embedding Snowflake UI within Salesforce through Canvas App:

Description: Embed Snowflake's UI directly into Salesforce account pages using a Canvas App. The Canvas App can provide direct interaction with Snowflake within the Salesforce interface.

Pros: Provides direct access to Snowflake features, minimizes data synchronization challenges.

Cons: Requires a deeper understanding of Snowflake's UI interaction, potential security considerations.

6. Implementing Scheduled Batch Jobs with Bulk API:

Description: Use scheduled batch jobs in conjunction with the Bulk API to periodically pull data from Snowflake and store it in Salesforce, displaying it on account pages.

Pros: Can be efficient for large data volumes, allows control over synchronization frequency.

Cons: Not real-time, potential complexity in handling data changes and synchronization.

Each of these approaches addresses various aspects of the user stories, ranging from data retrieval and security to error handling. The optimal solution would likely be a hybrid approach, carefully tailored to the specific needs, data volume, real-time requirements, and budget constraints of the project. Always consider testing the chosen approach thoroughly, ensuring that it aligns with the overall architectural best practices within the Salesforce ecosystem.

Considering an assortment of ways to implement an epic and associated user stories is vital to the decision-making process, enhancing the quality, scalability, and sustainability of the chosen solution. Examining multiple approaches fosters a comprehensive understanding of the problem domain. By comparing different methods, the development team gains insight into the intricacies of the requirements and possible hidden challenges or dependencies. This broader perspective empowers the team to design a solution closely aligned with current and anticipated future needs.

Reviewing various options often brings to light unique edge cases or scenarios that might have otherwise gone unnoticed. Discussing these situations in the context of different approaches allows the team to determine how best to handle them, reducing the risk of unforeseen issues. Furthermore, a structured review of different solutions fosters collaboration among team members, encouraging open dialogue and ensuring that various viewpoints and expertise are considered. This process leads to a more robust decision and promotes innovative and creative problem-solving.

Presenting multiple options, along with their pros and cons, to stakeholders demonstrates a thoughtful approach, reassures them that the chosen path has been carefully considered, and builds trust. It helps ensure alignment across the entire project team, including both technical and business stakeholders. Moreover, understanding why certain options were rejected can guide decisions in later stages of the project or even in subsequent projects, facilitating future development and maintenance.

The careful examination of different approaches also enhances risk management by evaluating the risks associated with each option. Understanding these risks upfront allows the team to make more informed decisions and mitigate potential issues. A thorough exploration of possible ways to implement epics and user stories, even those that may be rejected, is a strategic approach that fosters collaboration, encourages innovation, ensures alignment, and enhances the overall quality of the chosen solution. This reflective and methodical process is key to success in complex Salesforce development projects.

#1 – Using Salesforce Connect with Custom Adapter

Option *#1*, although presenting a reasonable and potentially effective solution, carries an underlying challenge that places it in the "no" column for the current phase of the project. The implementation of this option requires the purchase of additional licenses, a process that could introduce delays in the project timeline. Acquiring these licenses necessitates navigating through procurement, budget approval, and possibly legal review, all of which can extend the lead time significantly. In the context of a project where timeliness is paramount, this delay becomes a critical concern. While this option may still hold value for future considerations or in different scenarios, the immediate implications on schedule make it unsuitable for the current requirements and objectives. Therefore, the decision to set aside option *#1* in favor of alternatives that align more closely with the existing timeline and resources is a strategic and practical move.

#2 – Utilizing Custom Apex Classes and Callouts

Option *#2*, despite demanding more work upfront and necessitating a careful design to ensure scalability, offers a set of benefits that make it a compelling choice for the current project needs. Most notably, it does not require the acquisition of any additional Salesforce licenses, a factor that enhances its attractiveness by eliminating potential procurement delays. Furthermore, this approach is the most flexible option, providing a tailored solution that can be closely aligned with specific project requirements. Its capability to deliver near-real-time results adds another layer of appeal. The confidence in this option is further bolstered by the development team's prior experience in building custom integrations. Their hands-on expertise with similar projects not only assures that the implementation can be executed well but also fosters optimism that it can be accomplished quickly. This combination of flexibility, speed, and in-house expertise makes option *#2* a strong contender that aligns with both the technical needs and the organizational capabilities.

#3 – Integration Through Middleware (for example, MuleSoft)

Option #3 stands out as a likely candidate for selection in specific circumstances, most notably if there was already an established MuleSoft infrastructure in place along with an experienced Mulesoft team to manage and implement the integration. The synergies between MuleSoft and Salesforce could offer substantial benefits and streamline the development process. However, the present challenges around procuring MuleSoft and the time and resources required to build or hire a skilled MuleSoft team render this option impractical at this point in time. The complexities and potential delays associated with these challenges firmly place this option in the "no" column for the immediate project. Nevertheless, the strengths of this option should not be overlooked, and it would be prudent for the team to carefully review the "pros" associated with it. Insights gained from this analysis may prove valuable, allowing the team to incorporate elements of this approach where practical into the option eventually selected, thereby enhancing the final solution's robustness and functionality.

#4 – Utilizing Salesforce Einstein Analytics with Direct Data Connectors

Option #4, similar to option #3, presents certain procurement and team skill challenges that must be carefully considered. While this approach might align with certain aspects of the project, it also exhibits an overlap with the existing capabilities of Snowflake. Given its narrow focus on specific capabilities, it potentially limits future flexibility and optimization. If the required software was already procured and the team possessed the necessary skills, this option could be a viable consideration. However, in the current scenario, it doesn't align with the immediate project needs and timelines. It's essential to recognize the insights and perspectives that have shaped the support for this approach, and although it is not the optimal path at this moment, the considerations behind it are valued. By maintaining open communication and collaborative exploration of these ideas, we ensure that all stakeholders feel heard and engaged, fostering a united vision and a shared commitment to selecting the best solution for the project's success.

#5 – Embedding Snowflake UI within Salesforce through Canvas App

Option #5, while an intriguing suggestion, bears some characteristics that render it less suitable for the current project requirements. The approach is somewhat dated, reflecting methodologies that may not align with contemporary best practices within the Salesforce and Snowflake ecosystems. Implementing this option would likely introduce undue complexity around data movement and challenges in maintaining a consistent UI look and feel. Additionally, neither the Salesforce nor Snowflake teams have experience or expertise in this specific methodology, which could lead to potential delays and inefficiencies in the development process. Given these factors, although the idea is appreciated and its exploration reflects a healthy curiosity toward innovation, it falls into the "no" column for this particular project. It serves as a reminder that sometimes a thoughtful examination of various approaches can lead to a more deliberate and confident selection of the best-fit solution.

#6 – Implementing Scheduled Batch Jobs with Bulk API

Option #6, while offering certain advantages in some contexts, does not align with the specific requirement for the integration to be real-time in the current project. Choosing this approach would entail storing Snowflake data in Salesforce, thereby introducing synchronization challenges that could complicate the overall architecture and increase maintenance overhead. Managing such synchronization would diverge from the desired real-time interaction between Salesforce and Snowflake and could lead to inconsistencies in data representation. While there may indeed be scenarios where this option should be carefully considered, implementing it would necessitate a comprehensive revisit of the user stories and underlying requirements. Given the current project scope and the alignment sought between technical capabilities and business needs, this option is categorized as a "no" for the moment. The assessment serves as a reminder of the importance of closely matching implementation strategies with clearly defined project goals and requirements.

Conclusion – Integrating Salesforce and Snowflake

For the purposes of this book, #6 is a "no" because it does not meet the requirements. #5, #4, #3, and #1 are "no" due to a mix of procurement and team skills. This leaves #2, *Utilizing Custom Apex Classes and Callouts*.

ChatGPT is an assistant and accelerator. Rather than the development team members each reading the epic and thinking about the solution options (and likely not thinking about all of them, and then attending a meeting to espouse a particular solution based on an incomplete consideration of the options and the related pros and cons), ChatGPT was able to quickly provide the information needed for the team to align with a decision.

While option #2 is presently being recommended for its favorable balance of procurement concerns, licensing costs, existing team skills, and adaptability to future changes, it's essential to acknowledge the inherent strengths of option #3. The merit of option #3 lies in its potential scalability, modularity, and efficiency, possibly facilitated by existing middleware such as MuleSoft, if it were in place. By conducting a deep dive into these attributes, the development team can discover ways to enhance the #2-based solution. For instance, the modular design principles and integration patterns that are typically inherent in option #3 could be emulated within option #2. This might involve creating reusable integration components and employing best practices that allow for streamlined data exchange and transformation. The analytical comparison between options #2 and #3 exemplifies the importance of evaluating multiple solutions during the design phase. It's not merely about finding a solution that will work but synthesizing the best solution by carefully considering and potentially blending the strengths of various alternatives. Such a thoughtful approach ensures a more resilient and adaptable system that's well-aligned with both current needs and potential future evolutions.

Brainstorming with ChatGPT

Using ChatGPT to brainstorm solutions can be used from very basic epics and user stories (for example, *Epic 6, User Story 6.3: Implement Functionality to Invoke Report*, in *Appendix A – Case Study*), all the way through to a high-level requirement pending analysis and the epics and user stories being written (for example, *Create a React frontend that uses GraphQL to view and edit Salesforce Accounts*).

Prompt	I have a project requirement to create a React frontend that uses GraphQL to view and edit Salesforce accounts. What are the major components of this work?
Response	Certainly! The project requirement to create a React frontend that utilizes GraphQL to view and edit Salesforce accounts is complex and involves multiple major components. Below, I outline the components and possible options for developing each one.
	1. React Frontend Development:
	Components:
	UI Design: Creation of the interface where users will interact with the accounts.
	State Management: Managing the state of the application to keep track of changes in real time.
	Error Handling: Providing feedback to the users in case of failures or validation errors.
	Options:
	UI Libraries: Consider using Material UI, Ant Design, or other UI frameworks for consistent styling.
	State Management Libraries: Utilize Redux, MobX, or React's Context API for state management.
	Validation Libraries: Employ libraries such as Formik or Yup for form validation.
	2. GraphQL Integration:
	Components:
	Schema Definition: Defining the types, queries, and mutations that map to the Salesforce accounts.
	Resolver Implementation: Creating functions to handle fetching and manipulation of data.
	Authentication and Authorization: Ensuring only authorized users can access and modify the data.

Options:

Server: Use Apollo Server, Express-GraphQL, or other server middleware compatible with GraphQL.

Client-side Libraries: Employ Apollo Client or Relay Modern for client-side GraphQL management.

Security Measures: Implement JWT or OAuth 2.0 for authentication.

3. Salesforce Integration:

Components:

Connection Configuration: Setting up secure communication between Salesforce and your application.

Data Mapping and Transformation: Ensuring the data structure aligns with Salesforce objects.

Batch Operations: Handling bulk view and edit operations.

Options:

Connection Libraries: Utilize JSforce or other Salesforce SDKs for establishing a connection.

Middleware Integration: Consider using MuleSoft or other middleware if complex transformations are needed.

API Utilization: Leverage Salesforce REST or Bulk APIs as per the requirement.

4. Testing and Deployment:

Components:

Unit and Integration Testing: Validating individual components and their interactions.

Deployment Automation: Continuous deployment for releasing changes smoothly.

Options:

Testing Frameworks: Use Jest, the React Testing Library, or other testing frameworks.

Deployment Tools: Consider deploying with Docker, Kubernetes, or using a Platform-as-a-Service solution such as Heroku.

	The development of each component should align with the business goals and constraints such as budget, timeline, and expertise. Leveraging existing libraries and tools can accelerate development while ensuring that a scalable, secure, and maintainable solution is crafted. Collaboration with stakeholders and cross-functional teams throughout the development will ensure alignment with strategic objectives and a successful project execution.

Using ChatGPT at this level should only be done by architects and senior developers who know their way around Salesforce. This is because when working at this level, ChatGPT may miss the correct answer. In this case, ChatGPT missed that there is a Salesforce GraphQL API. This may be because ChatGPT's training material is out of date or because there is simply very little on the internet about the Salesforce GraphQL API.

Let's ask ChatGPT a direct question about the Salesforce GraphQL API:

Prompt	Salesforce has a GraphQL API. How much do you know about it?
Response	As of my knowledge cut-off date in September 2021, Salesforce does not offer a native GraphQL API out of the box. Instead, Salesforce primarily provides REST and SOAP APIs for external integration.
	Many developers have worked around this by implementing a GraphQL server that acts as an intermediary between Salesforce and the frontend application. The GraphQL server would be responsible for translating GraphQL queries into the appropriate SOQL (Salesforce Object Query Language) or API calls, and then translating the responses back into the GraphQL format.
	Such an implementation would typically involve defining GraphQL schemas that map to Salesforce objects, writing resolvers to handle the translation between GraphQL and Salesforce's APIs, and implementing appropriate authentication and authorization to ensure that access is properly controlled.
	Some third-party tools and libraries may offer GraphQL interfaces to Salesforce, and the specifics of these solutions would depend on the particular technologies involved. It's also worth considering that the landscape of available technologies is constantly evolving, and there may be new developments related to GraphQL and Salesforce since my last update.
	If Salesforce has introduced a native GraphQL API after my knowledge cut-off, I recommend referring to the official Salesforce documentation and community forums for the most accurate and up-to-date information. This will provide guidance on how to use the GraphQL API with Salesforce, including details on authentication, query execution, error handling, and more.

There is the answer – because the Salesforce GraphQL API was announced in 2022 and **Generally Available** (**GA**) in the Salesforce Winter '23 release, ChatGPT can't help, at least, at the time this chapter was written.

ChatGPT, and similar language models, are powered by **deep learning** (**DL**) algorithms that have been trained on vast datasets comprising a wide variety of textual content. These models are not merely stringing words together but are identifying complex patterns, relationships, and structures within the language. By analyzing countless examples of human-written text, the models develop an intricate understanding of syntax, semantics, and context. This enables them to generate coherent and contextually relevant text that aligns with human-like expression. The training process involves fine-tuning weights in the **neural network** (**NN**) to minimize errors in predictions, thereby facilitating the model's ability to generate text that resembles the way humans communicate. It is this robust interaction with language data, and the sophisticated architecture that interprets it, that empowers ChatGPT to respond to a wide array of prompts with remarkable fluency and relevance.

In a philosophical context, an entity that knows everything about everything is often referred to as "omniscient." In the realm of current technology, there's no existing entity that can truly claim omniscience, as even the most advanced **artificial intelligence** (**AI**) systems have limitations in their understanding and knowledge. While the term might be used metaphorically in some contexts, it's essential to recognize the boundaries and constraints of human-made systems and the ongoing exploration and learning that defines the field of **generative AI** (**GenAI**) and, specifically, ChatGPT.

Scaling solutions

Salesforce's robust platform provides scalable solutions for configuration changes, enabling virtually limitless growth. However, when it comes to writing custom Apex code, the onus is on developers to ensure that the code can handle the expected level of growth without hitting the platform's governor limits. These are built-in restrictions (for example, Governor Limits) that prevent undue resource consumption, and they present a unique challenge in the Salesforce ecosystem. Writing code that scales seamlessly requires a thoughtful approach that takes these limits into account. Typically, a project will document its anticipated growth and scaling needs, providing a clear benchmark for developers. This guidance allows them to architect solutions that not only meet current demands but also can gracefully handle projected increases in data volume and user interaction. By aligning Apex code with these growth expectations, developers can craft efficient, scalable solutions that harmonize with Salesforce's inherent strengths while avoiding the pitfalls of governor limits.

ChatGPT can serve as an instrumental asset in analyzing a user story requiring Apex code development within Salesforce, focusing on enhancing scalability. Through a comprehensive understanding of the underlying business needs, including expected levels of growth, it can grasp the core requirements of the user story. This insight becomes the foundation for suggesting appropriate coding practices tailored for scalability.

Instead of merely evaluating existing solutions, ChatGPT can actively engage in the development process, functioning akin to a "pair programmer." As the code is being developed, ChatGPT can continuously assess it, pinpointing areas where scalability might become a concern and providing real-time guidance. This partnership ensures that scalability principles are embedded right from the start, preventing potential bottlenecks.

Leveraging its vast knowledge of Salesforce development, ChatGPT can also propose suitable design patterns, best practices, efficient query structures, bulk processing techniques, asynchronous processing methods, and more. It can even generate customized code snippets that align with scalable development practices, serving as a starting point for the developers.

This collaboration extends beyond initial development, as ChatGPT continues to assist in code reviews and refinements throughout the life cycle of the project. By guiding the creation of relevant test cases targeting scalability, it ensures that the code is robust enough to withstand the pressures of expected growth.

ChatGPT's multifaceted approach transcends mere suggestion, transforming into a continuous engagement that resonates with both immediate needs and future growth. It acts as a virtual mentor, guiding developers to craft scalable solutions and ensuring alignment with Salesforce's scalability principles. By incorporating elements of analysis, real-time collaboration, and iterative refinement, ChatGPT fosters a development environment that is not only innovative but robust and future-proof.

Let's take a user story from *Appendix A – Case Study* and ask ChatGPT to analyze it for potential scaling issues:

Prompt	The user story below is going to include a Call Out. As the number of Salesforce users increases to the point that there will be multiple simultaneous calls from Salesforce to Snowflake, and the load on Snowflake may result in it taking seconds or more for a response, what governor limits should be front of mind and managed in the code?
	Epic 1: Integration of Snowflake Data with Salesforce Account Pages
	User Story 1.2: Call Out to Request Specific Account Information from Snowflake
	• **Goal**: As a Sales Rep, I want Salesforce to make a Call Out to Snowflake to request specific information related to a Salesforce account when I view the account page, providing me with timely insights.
	• **Acceptance Criteria**:
	• **Given** a Sales Rep is viewing a specific Salesforce account page and a secure connection to Snowflake is established,
	• **When** the page is loaded,
	• **Then** the system should make a Call Out to Snowflake to retrieve the relevant information for that specific account.
	• **Notes**: Specific, focused on the Call Out mechanism, and meets INVEST criteria.

| Response | The described user story, focusing on making a Call Out to Snowflake for specific account information, will indeed introduce specific challenges as the number of simultaneous Salesforce users increases. Managing Salesforce's governor limits will be paramount in ensuring that the code scales with the user load and handles any latency from Snowflake.

Here are the governor limits that should be front of mind and managed within the code:

1. Concurrent Remote Call Outs: Salesforce imposes a limit on the number of simultaneous remote callouts that can be made at a given time. This needs to be carefully managed, especially as multiple users might trigger callouts simultaneously.

2. HTTP Request Timeout: The standard timeout for HTTP callouts is 120 seconds. Given that the load on Snowflake might result in delays, it's essential to manage this timeout to prevent user experience issues. You may choose to modify this based on expected response times from Snowflake, but you'll need to strike the right balance.

3. Maximum CPU Time: Salesforce enforces limits on the maximum CPU time on the Salesforce servers. Writing efficient code to manage callouts will be crucial to ensure that this limit is not exceeded, especially under high load.

4. Asynchronous Callout Limits: If the integration utilizes asynchronous callouts to manage potential delays from Snowflake, keep in mind the limit on the number of future method invocations or Batch Apex jobs.

5. Total Number of Callouts in a Transaction: There is a limit on the number of callouts that can be made within a single transaction. This may need to be considered if multiple callouts are made within the same context.

6. Queueable Apex: If using Queueable Apex to manage callouts, the number of jobs added to the queue must be monitored to stay within limits.

7. Handling Errors and Retries: Implementing proper error handling and retry logic will be necessary to manage potential failures due to hitting these or other unforeseen limits.

These constraints necessitate a careful, thoughtful design that can effectively scale with the expected load. It might be beneficial to consider bulkifying callouts where applicable and utilizing asynchronous patterns to manage the load and potential delays from Snowflake. By keeping these governor limits in mind and building around them, a scalable and robust integration can be achieved. |
|---|---|

ChatGPT can be used to do a deep dive into each of these issues and provide guidance on how to navigate each of these limits. In scenarios where the needed prompt is not clear, ChatGPT can help refine the needed prompt. When getting into complex issues, it is important to not just blindly run with what ChatGPT produces. For example, in the preceding response, the responses around concurrent remote callouts and asynchronous callout limits both make assumptions about implementation and blend conditions together. This is the same thing an experienced developer might do when asked the same questions, but it is important to use ChatGPT to dive into the details of an answer if it is at all unclear.

Prompt	How can ChatGPT help understand and address the first limit mentioned previously: Concurrent Remote Call Outs?
Response	ChatGPT can serve as a valuable tool for Salesforce developers, providing insights and guidance on managing each of the mentioned governor limits. By engaging with ChatGPT, developers can explore potential challenges, solutions, and best practices related to these limits. This interactive exploration allows for a more comprehensive understanding of how to build scalable Apex code tailored to specific scenarios such as the Call Out to Snowflake.
	For example, if a developer wants to delve into the concurrent remote callouts limit, they might pose a prompt to ChatGPT such as:
	"Given a Salesforce integration that involves multiple simultaneous callouts to Snowflake for retrieving specific account information, I'm concerned about hitting the limit on concurrent remote callouts. Can you provide details on how this limit might impact my user story and suggest strategies to avoid exceeding this limit while maintaining performance?"
	Through this prompt, ChatGPT can provide detailed insights into how the concurrent remote callout limit operates, the potential consequences of hitting this limit, and various solutions that can be implemented. Solutions might include implementing a queuing mechanism, optimizing callout patterns, or utilizing asynchronous callouts with Queueable Apex. By leveraging ChatGPT's capabilities, developers have a resource to consult, helping them navigate the complex landscape of governor limits and craft code that aligns with Salesforce's best practices.

ChatGPT is trying to respond to the prompt but is still blending limits in the response. This exchange makes the point that as much as ChatGPT is a fantastic and knowledgeable assistant, validating answers, especially at the design level, is a good idea. Taking a minute to look through Salesforce documentation to ensure a complete understanding of the applicable limits is a very good idea.

While things are getting a little bit meta here, let's try the prompt ChatGPT suggested:

Prompt	Given a Salesforce integration that involves multiple simultaneous callouts to Snowflake for retrieving specific account information, I'm concerned about hitting the limit on concurrent remote callouts. Can you provide details on how this limit might impact my user story and suggest strategies to avoid exceeding this limit while maintaining performance?

Response	Certainly! The concurrent remote callouts limit in Salesforce is a critical consideration when dealing with simultaneous connections to external systems such as Snowflake. Hitting this limit can lead to throttling, delays, and even failed transactions, thereby affecting the user experience:
	1. Impact on Your User Story: In your specific scenario, where Salesforce is making multiple simultaneous callouts to Snowflake to fetch account information, exceeding the concurrent remote callouts limit might result in some of the callouts being delayed or failing. Since your user story requires timely retrieval of information, hitting this limit can cause inconsistencies, longer loading times, or partial failures, depending on how the integration is structured.
	2. Strategies to Avoid Exceeding the Limit:
	Implement Queuing Mechanism: By introducing a queueing system, you can control the number of concurrent callouts, ensuring that they are processed in a controlled manner. This prevents overwhelming the system while still allowing for efficient processing.
	Utilize Asynchronous Callouts with Queueable Apex: By handling callouts asynchronously, you can better manage the load on the system, distributing callouts over time and thus avoiding hitting the governor limit.
	Optimize Callout Patterns: If possible, try to group requests or minimize the number of simultaneous callouts required. Using Bulk API calls or reducing the frequency of callouts based on user behavior or needs can help in maintaining performance without breaching the limit.
	Monitoring and Error Handling: Implement robust logging, monitoring, and error-handling mechanisms to detect potential issues early and handle them gracefully. This not only provides a safeguard against hitting the limit but also ensures a better user experience if something goes wrong.
	3. Performance Considerations: While implementing these strategies, it's crucial to balance the need to stay within governor limits with the requirement for real-time insights. Careful planning, testing, and ongoing monitoring will help in finding the right balance that meets the business requirements without sacrificing performance.
	By understanding the specific demands of your integration and applying these strategies, you can develop a robust and scalable solution that aligns with Salesforce's governor limits and delivers the desired user experience. It's always a good practice to conduct thorough testing under simulated load conditions to validate the chosen approach in the real-world context of your application.

While ChatGPT has made several good recommendations, it does feel like the solution is going down a rabbit hole of complexity. Before building a complex integration that still may not quickly display up-to-date Snowflake data on a near-real-time basis, it is worth reviewing the design options ChatGPT suggested and seeing if with a change in requirements, there is a straightforward way to tackle this work.

Prompt	This is getting complicated. Earlier in this thread, one of the options you recommended was: Implementing Scheduled Batch Jobs with Bulk API. While this would not meet the requirement for the Snowflake information to be real-time, would this approach avoid the limits you suggested might be a concern?
Response	Yes, implementing scheduled batch jobs with the Bulk API would indeed circumvent many of the governor limits that might concern simultaneous callouts, but it does introduce a trade-off in terms of real-time data retrieval.
	Here's how this approach works and the implications:
	1. Batch Processing: Scheduled batch jobs allow you to collect multiple requests and process them as a single bulk operation. By aggregating requests and making fewer overall callouts, you significantly reduce the risk of hitting the concurrent remote callouts limit.
	2. Time Delay: However, this approach inherently introduces a time delay as the system waits to accumulate a batch of requests before processing them. This means that the Snowflake data retrieval would no longer be real-time, as callouts would be made at scheduled intervals or once enough requests are queued.
	3. Suitability for Your Use Case: If real-time retrieval of Snowflake data is a core requirement for your user story, this approach may not be suitable. However, if there is flexibility in the timing requirements or if the data can be slightly delayed without affecting the overall user experience, this option might be a practical solution.
	4. Scalability: This approach also offers better scalability, as batching callouts allows for more efficient utilization of resources, making it easier to manage increased load as the number of users grows.
	5. Complexity: Implementing scheduled batch jobs does add a layer of complexity to the system. It requires careful design to ensure that the batching, scheduling, and data retrieval processes are aligned with the business needs and user expectations.
	In conclusion, while implementing scheduled batch jobs with the Bulk API would likely alleviate governor limit concerns, it's essential to weigh the trade-offs, particularly the shift away from real-time data access. If the real-time requirement can be relaxed, this approach offers an efficient and scalable solution that can be tailored to suit your specific integration needs. It would be prudent to collaborate with the stakeholders to discuss this option and ensure that it aligns with the overall objectives of the integration with Snowflake.

ChatGPT serves as a remarkable partner and accelerator in the various stages of project work, from ideation to execution. It can provide invaluable insights, generate code snippets, analyze user stories, and offer solutions to complex technical challenges. The synergy of machine learning (ML) and human expertise opens doors to rapid innovation, efficiency, and creativity.

However, when it comes to intricate design work, especially those instances where the project might require suggesting changes to the foundational requirements, the human touch of an experienced architect or senior developer remains essential. While ChatGPT can furnish extensive data, ideas, and technical possibilities, the nuanced understanding, judgment, empathy, and strategic vision required to align technology with business goals are uniquely human traits. An experienced architect will not only consider the technical dimensions but also the broader organizational context, stakeholders' needs, long-term scalability, and potential future directions. This ability to synthesize complex variables and navigate the often delicate terrain of changing requirements is something that, at least for the moment, requires the depth and breadth of human expertise. Therefore, while ChatGPT can be a vital tool in the arsenal of a development team, it complements rather than replaces the irreplaceable insight, wisdom, and guidance that seasoned human professionals bring to the table.

Summary

You've now enhanced your ability to harness ChatGPT for mapping out multiple Salesforce design strategies tailored to specific user stories. Throughout this chapter, you've absorbed the practice of critically weighing the strengths and limitations of each design option, underpinned by a thorough understanding of Salesforce's varied methodologies and best practices. You're now equipped to make more strategic decisions, with a keen eye on scalability, maintainability, and alignment with business objectives. The collaborative decision-making skills you've refined will serve you in fostering team consensus and driving effective, adaptable solutions within your Salesforce projects.

In the next chapter, you'll learn the art of dissecting and refining Salesforce features created by others. This chapter will guide you through the analytical process of understanding the purpose and mechanics of pre-existing Apex, teaching you how to assess its functionality and identify any underlying issues. You'll learn how to address these issues effectively, ensuring your Apex code adheres to best practices and serves business needs seamlessly. Additionally, the chapter will highlight the importance of clear commentary in code, offering strategies for documenting your work to aid in future maintenance and collaboration. By the end of this chapter, you'll be adept at leveraging ChatGPT to enhance your approach to interacting with and improving legacy Apex code, elevating your capacity for maintaining robust and efficient Salesforce applications.

Using ChatGPT for Salesforce Apex That Someone Else Wrote

Salesforce Apex represents a remarkable tool in the Salesforce ecosystem, allowing developers to write and execute robust and highly customizable server-side logic. As the powerhouse behind Salesforce's programmability, Apex offers the ability to create dynamic business logic, automate processes, and integrate with external systems. Unlike declarative automation tools such as flows, Apex provides a finer level of control, enabling developers to meet complex requirements that might be beyond the reach of out-of-the-box tools. Apex's expressive syntax, drawn from Java, makes it a potent instrument for solving intricate business problems.

However, the complexity of Apex is a double-edged sword. Writing Apex code that is effective, efficient, and maintainable demands a sound understanding of software development principles, the Salesforce platform's nuances, and the limitations enforced by Salesforce. Apex developers must be well versed in governor limits, which restrict the number of operations in a single transaction. Handling these constraints while ensuring that the code is optimized for performance becomes a significant challenge. Developers must think not just in terms of functionality but also about how the code will operate within the platform's limitations.

Moreover, the challenge escalates as Salesforce organizations grow in terms of data volumes, users, and usage. Apex code that may have initially been effective can quickly become a bottleneck as the scale increases. Writing Apex that scales requires careful consideration of query optimization, efficient memory utilization, and a deep understanding of how different parts of the Salesforce platform interact. It's not merely about getting the code to work; it's about building solutions that will continue to perform as demands on the system expand.

Adding to the complexity is the need to ensure code maintainability and compliance with best practices. With multiple developers working on a code base over time, poorly written or inadequately documented Apex can become a nightmare to understand and modify. Tools such as Salesforce Health Check can provide insights into optimization, but it often requires an experienced eye to discern where improvements can be made.

While Apex's power is undeniable, wielding it effectively requires a blend of technical skill, platform knowledge, and foresight to anticipate how solutions must adapt to future growth and change. Mastery of Apex is not merely about understanding syntax and structure but also involves a strategic approach to developing solutions that are efficient, scalable, maintainable, and aligned with the intricate nature of the Salesforce environment. It's a tool that enables enormous creativity and flexibility but demands respect for its complexity and potential pitfalls.

Salesforce's approach to innovation is characterized by its tri-annual release cycle, where three major updates are rolled out every year. These releases are a cornerstone of Salesforce's commitment to providing the latest features, improvements, and security enhancements. However, they're not optional for organizations; every Salesforce instance is automatically upgraded with each new release. While this ensures that all customers have access to the latest capabilities, it introduces a unique challenge, particularly concerning Apex code.

Apex code, being tightly integrated with the underlying Salesforce platform, can be affected by changes and updates introduced in these releases. What worked seamlessly in one version might suddenly become deprecated or behave differently in a subsequent release. This necessitates regular review and, often, refactoring of Apex code to ensure that it continues to function as intended. It's not just about keeping up with new features but also about ensuring that existing functionality doesn't break or degrade in performance. This continuous adaptation can require significant effort, especially in larger or more complex Salesforce implementations. It emphasizes the importance of following best practices, keeping abreast of upcoming changes, and adopting a proactive approach to managing Apex code. Developers must not only consider the immediate requirements but also how their code will adapt to the ever-evolving Salesforce landscape, maintaining a delicate balance between leveraging the latest platform capabilities and ensuring ongoing stability and performance.

It's a common trend in the software development world to describe new **object-oriented programming** (**OOP**) languages as "being like Java." This comparison serves as a shorthand to convey some essential characteristics of the language, such as syntax, structure, or fundamental concepts. However, when it comes to Salesforce Apex, this comparison can be somewhat misleading and overlooks several crucial aspects that set Apex apart from Java, particularly in the context of governor limits.

Governor limits in Apex add a significant layer of complexity that's unique to the Salesforce environment. These limits are system constraints that prevent individual users from monopolizing shared resources, thereby ensuring that the system performs efficiently and avoids any negative performance impact on others. For example, there are limits on the number of SOQL queries, DML statements, and CPU time that can be consumed in a single transaction. Unlike in Java, where developers usually have wide latitude to write code as they see fit, Apex developers must always keep these governor limits in mind.

Many Apex developers, especially those new to Salesforce, find themselves learning these concepts "on the job." The lack of awareness of governor limits often leads to writing code that, while it might function correctly in a narrow context or small dataset, could run into limits under real-world conditions or as the organization's data volume grows. This makes the code brittle and can lead to unexpected failures.

This often results in a need for refactoring, where the original code must be rewritten to make it more efficient and resilient against governor limits. Refactoring can be a complex and time-consuming process, requiring a deep understanding of both the business logic and the nuances of Apex itself.

The characterization of Apex as "being like Java" can be a double-edged sword. While it might make the language more approachable for those familiar with Java, it may also lead to misunderstandings and oversights regarding the specific challenges posed by governor limits. A better approach might be to recognize Apex as a language with its unique constraints and capabilities, one that requires careful attention to details that might not be immediately apparent to those coming from other OOP backgrounds. The learning curve might be steeper, but the reward is code that's not only functional but also robust and scalable within the Salesforce environment.

The complexity of Apex can turn troubleshooting or updating existing Apex code into a non-trivial task, often involving the subtle unraveling of unexpected side effects and intricate dependencies. This is where the power of ChatGPT truly shines. By providing immediate insights and guidance, ChatGPT can bridge the knowledge gap, aiding both novice and experienced developers in quickly navigating the complex landscape of someone else's Apex. Rather than spending days or even longer chasing subtle issues, developers can leverage ChatGPT to expedite the completion of assigned work, enhancing efficiency and the overall quality of the code.

Understanding existing Apex code

ChatGPT's ability to analyze uncommented and unclear Apex code is a testament to its sophisticated understanding of programming principles and syntax. Even when faced with code that lacks meaningful variable names, method names, or explanatory comments, ChatGPT can dissect the logic, structure, and operations within the code to provide an accurate description of what the code does. By identifying the sequence of method calls, the relationships between variables, and the computational processes involved, ChatGPT can elucidate the underlying functionality of the code. However, it is essential to recognize that the business context or the specific purpose behind the code may not be inferable from the code itself, especially when it is obfuscated or intentionally made unclear. Therefore, while ChatGPT can illuminate the technical aspects of what the code is doing, understanding why it does so may require additional context or information outside the code itself.

Prompt	What does this code do?

```
public class AppleBananaCherry {

    private static final Double CAT = 3958.8;

    private static Double DOG(Double a, Double b, Double
c, Double d) {
        Double E = toFish(c - a);
        Double F = toFish(d - b);
        Double G = Math.sin(E / 2) * Math.sin(E / 2) +
                    Math.cos(toFish(a)) * Math.
cos(toFish(c)) *
                    Math.sin(F / 2) * Math.sin(F / 2);
        Double H = 2 * Math.atan2(Math.sqrt(G), Math.
sqrt(1 - G));
        return CAT * H;
    }

    private static Double toFish(Double frog) {
        return frog * Math.PI / 180;
    }

    public static void elephantHorse(Id giraffe) {
        Account lion = [SELECT BillingLatitude,
BillingLongitude FROM Account WHERE Id = :giraffe LIMIT
1];

        Integer monkey = 0;
        Integer snake = 0;

        for (Contact zebra : [SELECT MailingLatitude,
MailingLongitude FROM Contact WHERE AccountId = :giraffe])
{
            Double panda = DOG(lion.BillingLatitude,
lion.BillingLongitude, zebra.MailingLatitude, zebra.
MailingLongitude);
            if (panda < 20) {
                monkey++;
            } else {
                snake++;
            }
        }
```

|  | ```
        System.debug('Output 1: ' + monkey);
        System.debug('Output 2: ' + snake);
    }
}
``` |
| **Response** | The given code defines a class `AppleBananaCherry` in Salesforce's Apex language that is responsible for calculating the geographical distances between an Account's address and the addresses of its associated Contacts.

Here's a summary of what the code does:

1. Constant Definition: The constant `CAT` is defined with the value 3958.8, which seems to be the radius of the Earth in miles, a common value used in distance calculations.

2. Distance Calculation (Haversine formula): The private method `DOG` calculates the geographical distance between two latitude-longitude pairs using the Haversine formula. The `toFish` method converts degrees to radians.

3. Main Processing Method (`elephantHorse`): This public method takes an Account ID (`giraffe`) as an input.

It retrieves the latitude and longitude of the billing address of the given Account.

It then queries all the Contacts associated with that Account and retrieves their mailing latitude and longitude.

For each Contact, it calls the `DOG` method to calculate the distance between the Account's billing address and the Contact's mailing address.

If the calculated distance is less than 20 miles, it increments a counter `monkey`.

If the distance is 20 miles or more, it increments a counter `snake`.

Finally, it outputs the values of `monkey` and `snake` to the debug log.

These represent the count of Contacts that are less than 20 miles and 20 miles or more from the Account's address, respectively.

The class, method, and variable names are intentionally misleading and have no clear relation to what the code is doing. Nonetheless, the actual code implements a specific geographical analysis based on the relationship between Accounts and Contacts in a Salesforce organization. |
```

At the time of this book's writing, there are constraints on the size of prompts that can be sent to ChatGPT. While ChatGPT is an incredibly powerful tool capable of handling a wide range of inquiries, including the analysis of Apex code, it's essential to recognize the limitations related to the length and complexity of the input.

For large Apex classes or modules, it may be necessary to approach the inquiry in a segmented manner, breaking down the analysis by module or even in smaller tranches of Apex code at a time. This approach ensures that the inquiry aligns with the prompt size restrictions and allows for a more detailed, focused analysis of specific sections of the code. Working incrementally in this way can be highly beneficial, particularly when dealing with complex logic or multiple interrelated components within the Salesforce ecosystem. It helps to maintain clarity, supports efficient troubleshooting, and allows developers to gain precise insights into the code's functionality and structure, thereby enhancing their understanding and ability to work with the code.

Identifying and rectifying issues in Apex code

ChatGPT's ability to analyze Apex code serves as a valuable asset in uncovering underlying issues that may not be immediately apparent. With a nuanced understanding of Salesforce best practices and governor limits, ChatGPT can inspect code, even when it appears to work in limited scenarios, and identify problematic areas that may lead to scalability and maintainability challenges. By assessing the code structure, query placement, handling of bulk operations, and other factors, ChatGPT can provide actionable insights and suggestions to refactor the code, aligning it with recommended design principles. This capability is not only beneficial for developers early in their Salesforce career but also serves as a second pair of eyes for experienced developers, ensuring that the code is robust and ready for real-world scenarios, irrespective of data volumes or complexity.

Prompt	Is there anything in this code that does not follow Salesforce best practices and that you would recommend rewriting? If so, please identify what is wrong, why it is wrong, and how it should be fixed.

```apex
public class AnalyzeContactDistance {

    private static final Double EARTH_RADIUS = 3958.8;

    private static Double calculateDistance(Double lat1,
Double lon1, Double lat2, Double lon2) {
        Double dLat = toRadians(lat2 - lat1);
        Double dLon = toRadians(lon2 - lon1);
        Double a = Math.sin(dLat / 2) * Math.sin(dLat / 2)
+
                     Math.cos(toRadians(lat1)) * Math.
cos(toRadians(lat2)) *
                     Math.sin(dLon / 2) * Math.sin(dLon /
2);
        Double c = 2 * Math.atan2(Math.sqrt(a), Math.
sqrt(1 - a));
        return EARTH_RADIUS * c;
    }

    private static Double toRadians(Double deg) {
        return deg * Math.PI / 180;
    }

    public static void analyzeAccountContacts(Id
accountId) {
        Account account = [SELECT BillingLatitude,
BillingLongitude FROM Account WHERE Id = :accountId LIMIT
1];

        // Below query is inside the loop, a classic
example of poor design leading to governor limit issues.
        List<Contact> contacts = [SELECT MailingLatitude,
MailingLongitude FROM Contact WHERE AccountId =
:accountId];

        Integer nearCount = 0;
        Integer farCount = 0;
```

```
                for (Contact contact : contacts) {
                    for(Contact innerContact : [SELECT
        MailingLatitude, MailingLongitude FROM Contact WHERE Id =
        :contact.Id]) {
                        Double distance =
        calculateDistance(account.BillingLatitude, account.
        BillingLongitude, innerContact.MailingLatitude,
        innerContact.MailingLongitude);
                        if (distance < 20) {
                            nearCount++;
                        } else {
                            farCount++;
                        }
                    }
                }

                System.debug('Nearby Contacts: ' + nearCount);
                System.debug('Distant Contacts: ' + farCount);
            }
        }
```

Response	This code contains several significant issues:
	SOQL Query Inside the Loop: The nested query inside the loop will likely cause the code to hit the governor limits on the number of SOQL queries in a single execution context as the data volumes increase. This is a classic anti-pattern in Salesforce development.
	Inefficient Code Structure: The redundant queries and the inefficient way the calculations are performed mean that this code is likely to perform poorly, even if it doesn't hit governor limits on very low data volumes.
	Lack of Bulk Processing Considerations: This code does not take bulk processing into account, which would make it unsuitable for handling large numbers of records simultaneously.
	Such code might pass initial testing with small data sets but is highly prone to failure in a production environment or as data volumes grow, leading to maintenance headaches and potential business disruptions. It exemplifies why understanding governor limits and following Salesforce best practices is essential.

Beyond merely identifying issues in Apex code, ChatGPT possesses the ability to rewrite the code to address identified problems, aligning with best practices and optimizing performance. This can be a valuable asset in accelerating the code refinement process and ensuring adherence to robust coding standards. However, it's essential to emphasize that the rewritten code provided by ChatGPT should not be blindly trusted or implemented without further scrutiny. While the assistance is based on recognized best practices and expert insights, it's crucial that the revised code undergoes a comprehensive testing cycle to ensure that it not only resolves the identified issues but also exhibits the correct desired behaviors. This testing ensures alignment with business requirements and functional specifications, avoiding unintended side effects that could result from the modifications. As with any development process, collaboration with the development team, quality assurance, and subject-matter experts remains paramount to validate and fine-tune the solution.

Let's have ChatGPT rewrite this code:

Prompt	Please rewrite this Apex to address all of the issues and ensure it follows Salesforce best practices and will scale. Please note any remaining issues that should be considered and captured to be addressed in the future.
Response	Certainly! Below is the revised Apex code, following Salesforce best practices to ensure scalability and robustness.

```
public class AnalyzeContactDistance {

    private static final Double EARTH_RADIUS = 3958.8;

    private static Double calculateDistance(Double
lat1, Double lon1, Double lat2, Double lon2) {
        Double dLat = toRadians(lat2 - lat1);
        Double dLon = toRadians(lon2 - lon1);
        Double a = Math.sin(dLat / 2) * Math.sin(dLat /
2) +
                    Math.cos(toRadians(lat1)) * Math.
cos(toRadians(lat2)) *
                    Math.sin(dLon / 2) * Math.sin(dLon /
2);
        Double c = 2 * Math.atan2(Math.sqrt(a), Math.
sqrt(1 - a));
        return EARTH_RADIUS * c;
    }

    private static Double toRadians(Double deg) {
        return deg * Math.PI / 180;
    }

    public static void analyzeAccountContacts(Id
accountId) {
        Account account = [SELECT BillingLatitude,
BillingLongitude FROM Account WHERE Id = :accountId
LIMIT 1];
        List<Contact> contacts = [SELECT
MailingLatitude, MailingLongitude FROM Contact WHERE
AccountId = :accountId]; // Removed SOQL from the loop
```

```
        Integer nearCount = 0;
        Integer farCount = 0;

        for (Contact contact : contacts) {
            Double distance =
calculateDistance(account.BillingLatitude, account.
BillingLongitude, contact.MailingLatitude, contact.
MailingLongitude);
            if (distance < 20) {
                nearCount++;
            } else {
                farCount++;
            }
        }

        System.debug('Nearby Contacts: ' + nearCount);
        System.debug('Distant Contacts: ' + farCount);
    }
}
```

By removing the inner SOQL query from the loop, the revised code avoids the governor limit issue related to query execution within loops. This ensures that the code is more scalable and better aligned with Salesforce's best practices.

Remaining Considerations:

The method is still accessing all contacts associated with the account without any limit, which might be a concern if there's a significant number of contacts. Implementing some form of pagination or filtering criteria might be necessary in the future, depending on the actual usage pattern and data volumes.

Unit testing the method will be essential to validate the correctness of the revised code. Proper testing should ensure that the method behaves as expected under various scenarios and with different data inputs, considering all possible edge cases.

The hardcoded distance threshold (20 miles) could be made configurable through custom settings or metadata, enhancing flexibility and maintainability.

The importance of having a complete set of test classes, encompassing both positive and negative cases, cannot be overstated in the context of Salesforce development. Positive test cases validate that the code behaves correctly under expected conditions and meets the defined requirements. Negative test cases, on the other hand, ensure that the code can handle unexpected situations, errors, and edge cases gracefully, without causing failures or unexpected behavior. When a class is restructured or rewritten, as we have done with the assistance of ChatGPT, it's crucial to create new test classes to reflect the modifications. These newly crafted test classes must thoroughly cover the updated functionality, including all possible pathways and logical branches within the code. By diligently testing both the "happy path" and potential failure modes, developers can build confidence that the revised code is robust, reliable, and aligned with Salesforce best practices. It establishes a strong foundation that not only ensures code quality but also facilitates future development and maintenance efforts.

Prompt	Please write positive and negative test classes for the rewritten AnalyzeContactDistance class. Include a test that has an account with 200 contacts.

Response	Certainly! Below are the positive and negative test classes for the rewritten `AnalyzeContactDistance` class. We'll begin by writing the test setup, then proceed with the positive and negative test scenarios. The test with 200 contacts will be included in the positive test scenario.
	
	Test Setup
	
	``
	@isTest
	private class AnalyzeContactDistanceTestSetup {
	    @TestSetup
	    static void setupTestData() {
	        Account acc = new Account(Name = 'Test
	Account', BillingLatitude = 37.7749, BillingLongitude =
	-122.4194);
	        insert acc;
	
	        List<Contact> contacts = new List<Contact>();
	        for (Integer i = 0; i < 200; i++) {
	            contacts.add(new Contact(LastName
	= 'Contact' + i, MailingLatitude = 37.7749,
	MailingLongitude = -122.4194, AccountId = acc.Id));
	        }
	        insert contacts;
	    }
	}
	``

Positive Test Class

```
@isTest
private class AnalyzeContactDistancePositiveTest {
    @isTest
    static void testAnalyzeAccountContacts() {
        Account testAccount = [SELECT Id FROM Account
WHERE Name = 'Test Account' LIMIT 1];
        Test.startTest();
        AnalyzeContactDistance.
analyzeAccountContacts(testAccount.Id);
        Test.stopTest();
        // You may include System.assert statements
here to validate expected outcomes.
        // This could include querying related records
and checking that the correct actions have been taken.
    }

    @isTest
    static void testAnalyzeAccountWith200Contacts() {
        Account testAccount = [SELECT Id FROM Account
WHERE Name = 'Test Account' LIMIT 1];
        Test.startTest();
        AnalyzeContactDistance.
analyzeAccountContacts(testAccount.Id);
        Test.stopTest();
        // Include specific assertions to validate the
behavior with 200 contacts.
    }
}
```

Negative Test Class

```
@isTest
private class AnalyzeContactDistanceNegativeTest {
    @isTest
    static void testAnalyzeAccountContactsWithNullId()
{
        Test.startTest();
        try {
            AnalyzeContactDistance.
analyzeAccountContacts(null);
            System.assert(false, 'Expected an exception
but none was thrown.');
        } catch (Exception e) {
            System.assertEquals('Some expected error
message', e.getMessage());
        }
        Test.stopTest();
    }
}
```

Note: Replace ``'Some expected error message'`` with the actual error message that should be thrown when passing a null `Id`. The assertions in the positive test classes would depend on the specific expectations of the rewritten `AnalyzeContactDistance` class, so you may need to customize these based on the revised logic and desired outcomes.

Having ChatGPT write Apex test classes ensures that new functionality meets the required standards and behaves as expected. Having ChatGPT add comments to existing Apex enhances the maintainability and understandability of the code base. With ChatGPT's capabilities, existing Apex code can be annotated with explanatory comments, providing clarity on complex logic and workflows. This shift not only aids in immediate comprehension for developers but also serves as vital documentation for future reference, ensuring that the intent and function of the code are transparent and accessible to anyone who works on it subsequently.

Adding and managing comments in Apex code

Commenting, including both inline comments and comment headers, has long been a vital aspect of software development. It serves as a roadmap, guiding developers through complex pieces of code, providing context for why a particular solution was implemented, and outlining the expected inputs and outputs of a function. Well-commented code can significantly reduce the learning curve for new team members and foster a smoother collaboration between developers working on the same code base.

ChatGPT demonstrates a noteworthy proficiency in interpreting and commenting code. By analyzing the code structure, variables, and logic flow, it can generate comments that offer insights into the code's functionality, even assisting in the automatic documentation of methods and classes.

However, the growing ability of tools such as ChatGPT to decipher uncommented code and explain its functionality on the fly does raise intriguing questions about the future of code commenting. As these tools become more sophisticated and accessible, the need for extensive commenting may diminish, shifting the focus to writing clean and self-explanatory code. This doesn't negate the value of commenting, especially for intricate logic or areas where business rules and context are not evident from the code itself, but it does signify a potential evolution in how developers approach code readability. The "autumn of commenting code" may indeed be upon us, yet the wisdom and clarity imparted through well-placed comments will likely retain their value in ensuring maintainable and comprehensible code.

Prompt	Please add a complete set of comments to the AnalyzeContactDistance class, including a full comment header and inline comments.

Response	Certainly! Here's the `AnalyzeContactDistance` class with a complete set of comments, including a full comment header and inline comments to describe the purpose and functioning of the class, methods, and key logic.
	
	``
	/**
	 * Class: AnalyzeContactDistance
	 * Description: This class is designed to analyze the geographical distance between an Account's address and its associated Contacts' addresses.
	 * It calculates the number of Contacts that are less than 20 miles from the Account's address and the number that are more than 20 miles away.
	 *
	 * Usage:
	 * AnalyzeContactDistance.analyzeAccountContacts(accountId);
	 *
	 * Constants:
	 * EARTH_RADIUS - Radius of the earth used in distance calculation (in miles).
	 */
	public class AnalyzeContactDistance {
	
	    private static final Double EARTH_RADIUS = 3958.8; // Earth's radius in miles
	``

```
/**
 * Method: calculateDistance
 * Description: Calculates the distance between two
geographical points using the Haversine formula.
 * Parameters: lat1, lon1, lat2, lon2 -
Geographical coordinates of two locations.
 * Returns: Distance in miles between the two
locations.
 */
private static Double calculateDistance(Double
lat1, Double lon1, Double lat2, Double lon2) {
    Double dLat = toRadians(lat2 - lat1); //
Difference in latitude converted to radians
    Double dLon = toRadians(lon2 - lon1); //
Difference in longitude converted to radians
    Double a = Math.sin(dLat / 2) * Math.sin(dLat /
2) +
                Math.cos(toRadians(lat1)) * Math.
cos(toRadians(lat2)) *
                Math.sin(dLon / 2) * Math.sin(dLon /
2); // Haversine formula part
    Double c = 2 * Math.atan2(Math.sqrt(a), Math.
sqrt(1 - a)); // Inverse Haversine formula
    return EARTH_RADIUS * c; // Return distance
}


/**
 * Method: toRadians
 * Description: Converts degrees to radians.
 * Parameter: deg - Angle in degrees.
 * Returns: Angle in radians.
 */
private static Double toRadians(Double deg) {
    return deg * Math.PI / 180;
}


/**
 * Method: analyzeAccountContacts
 * Description: Analyzes the distance between an
Account's address and its Contacts' addresses,
 * categorizing them into two groups based on
whether they are closer or farther than 20 miles from
the Account.
 * Parameter: accountId - The ID of the Account to
analyze.
```

```
     * Side Effects: Logs the counts of nearby and
distant Contacts.
     */
    public static void analyzeAccountContacts(Id
accountId) {
        Account account = [SELECT BillingLatitude,
BillingLongitude FROM Account WHERE Id = :accountId
LIMIT 1]; // Retrieve Account

        List<Contact> contacts = [SELECT
MailingLatitude, MailingLongitude FROM Contact WHERE
AccountId = :accountId]; // Retrieve Contacts

        Integer nearCount = 0; // Count for Contacts
near the Account
        Integer farCount = 0; // Count for Contacts far
from the Account

        // Loop through Contacts to calculate distance
        for (Contact contact : contacts) {
            Double distance =
calculateDistance(account.BillingLatitude, account.
BillingLongitude, contact.MailingLatitude, contact.
MailingLongitude);
            if (distance < 20) {
                nearCount++; // Increment near count if
distance is less than 20 miles
            } else {
                farCount++; // Increment far count if
distance is 20 miles or more
            }
        }

        System.debug('Nearby Contacts: ' + nearCount);
// Log the count of nearby Contacts
        System.debug('Distant Contacts: ' + farCount);
// Log the count of distant Contacts
    }
}
```

The above comments provide a comprehensive overview of the class and its functionality, making the code more readable and maintainable for future developers.

The integration of ChatGPT in the Salesforce ecosystem represents a significant advancement in the way existing Apex code can be analyzed, improved, and maintained. By leveraging the capabilities of ChatGPT, developers can quickly dissect and understand complex, uncommented code, identifying areas that may require optimization or alignment with best practices. The analysis isn't just limited to identification; it also extends to offering actionable insights and guidelines on how to refactor the code to enhance performance, scalability, and readability. Such an approach accelerates the code review process, ensuring a higher code-quality standard.

The ability of ChatGPT to rewrite existing Apex to improve its efficiency goes beyond mere syntactic adjustments; it focuses on strategic code restructuring that aligns with Salesforce best practices and architectural principles. This rewriting ability supports developers in handling legacy code and ensures that even with increased data volumes, the code remains robust and compliant. While the rewritten code must still undergo rigorous testing to confirm its functionality, this AI-powered assistance marks a significant shift in how development and maintenance tasks are approached.

Moreover, the skill of ChatGPT to autogenerate comprehensive comments, both inline and headers, introduces a nuanced perspective on code documentation. While comments have always been crucial in understanding the intent and functioning of the code, ChatGPT's ability to make sense of uncommented code and provide clear explanations brings a new dynamism to code maintenance. It suggests a future where maintaining existing code is no longer a tedious or time-consuming task, but an efficient, AI-assisted process that emphasizes collaboration, learning, and constant improvement. This paradigm shift heralds an exciting era where human expertise and AI coalesce to elevate software development to new heights.

Summary

This chapter shared practical knowledge on understanding and updating existing Salesforce Apex code. This knowledge includes dissecting the functionality of existing Apex code, gaining insights into its intended operations and its role within the Salesforce org, strategies for identifying issues in the code, and outlining systematic approaches for remediation and optimization. Moreover, it underscored the significance of commenting in Apex code, teaching readers how to use ChatGPT to craft clear and purposeful comments that facilitate easier maintenance and foster better collaboration among development teams. With these skills, developers will be better prepared to ensure the continuity of business processes through effective management of legacy Apex code.

The next chapter begins an exploration of using ChatGPT for Salesforce Apex development. It begins by introducing the robust features of Apex and the supportive role ChatGPT can play in the development life cycle. The chapter then guides the reader through the nuances of utilizing ChatGPT to craft Apex code, emphasizing the tool's ability to provide coding assistance and enforce best practices. Error handling and debugging are also covered, with ChatGPT serving as a resource for quick error identification and resolution. Lastly, the chapter emphasizes the importance of staying abreast of best practices in Apex coding, positioning ChatGPT as a valuable ally in continuous learning and skill enhancement for developers.

6
Using ChatGPT
for Salesforce Apex

Apex is a proprietary programming language provided by Salesforce that's designed to allow developers to execute flow and transaction control statements on Salesforce servers. It enables developers to add business logic to system events, such as button clicks or updates, and create custom logic through Apex classes, batch jobs, and schedulable classes. Apex is tightly integrated with Salesforce's data model and is optimized for building complex business processes that interact with the Salesforce application.

ChatGPT offers significant value in this landscape, particularly in Apex development, by assisting with a wide range of activities, from code generation to debugging and even architectural design considerations. For instance, ChatGPT can guide a developer through best practices in writing Apex classes, including error handling, governor limits, and test class creation. It can analyze existing code for readability, complexity, and performance issues, offering refactoring suggestions for better maintainability and efficiency. ChatGPT can even assist in designing scalable and reusable patterns for your Apex classes to follow, ensuring you're not just solving the problem at hand but building a long-lasting solution.

By leveraging the capabilities of ChatGPT, developers can accelerate their development cycle, ensuring that they're following best practices while also catching potential issues before they become major problems. This not only makes the development process more efficient but also helps in producing higher quality, more reliable Apex code. This unique blend of functionalities makes ChatGPT an invaluable tool for anyone looking to master Apex development in Salesforce, whether they're newcomers trying to learn the ropes or experienced developers looking for optimized solutions.

Crafting high-quality Apex code in Salesforce presents its own set of unique challenges that developers need to address for robust and maintainable solutions. One of the most common issues is dealing with governor limits. Salesforce imposes various types of governor limits, such as SOQL query limits, DML statement limits, and CPU time limits, to ensure that no single tenant monopolizes shared resources in the multi-tenant environment. The challenge here is to write efficient and bulk-safe code that can perform the required tasks within these constraints, which often requires careful planning, bulkification of code, and the use of patterns such as batch and queueable Apex.

Another major challenge is ensuring robust security measures in Apex code. Apex runs in system mode, meaning it doesn't inherently respect the object and field-level permissions assigned to the running user. Developers have to explicitly check for CRUD, FLS, and sharing rules, especially when the code is intended to run in a user context. Ensuring that the code adheres to Salesforce's security model, while still being functional and efficient, requires a nuanced understanding of both Apex and the platform's security mechanisms.

Lastly, testing and debugging can be particularly challenging. Apex requires a minimum code coverage of 75% to be deployable, but achieving high code coverage is not synonymous with having well-tested code. Developers often face the challenge of writing meaningful test cases that cover various scenarios and edge cases, including bulk operations, partial failures, and so on. Debugging tools for Apex are improving but still have some way to go in terms of providing the kind of real-time, interactive debugging features available in other languages. These factors make the testing and debugging process more cumbersome and time-consuming than it could be otherwise.

Crafting Apex code with ChatGPT's guidance

As discussed previously, crafting Apex code can be a complex and nuanced task that requires not only a deep understanding of Salesforce's capabilities but also a keen eye for best practices and efficient problem-solving. Whether you're creating triggers, batch classes, or general utility classes, the quality of your Apex code directly impacts the robustness and maintainability of your Salesforce solution. This is where ChatGPT can offer unparalleled guidance.

Given that Salesforce's drag-and-drop functionalities are continuously improving, the following user story might be able to be addressed without Apex code. However, as a practical example, we can consider the following user story:

User story – *automate account review notifications*

- **Goal**: As a sales manager, I want to automatically send email notifications to account owners for accounts that have not been updated in the last 30 days

- **Acceptance criteria**:

  - **Given** an account has not been updated in the last 30 days

  - **When** the nightly batch job runs

  - **Then** the account owner should receive an email notification reminding them to review and update the account details

This user story would require an Apex Batch class to query all accounts and check their last modified date, then send an email to those that meet the criteria. You'd also probably want a custom object or field to log the notifications that are sent:

Prompt	Could you please create the Apex code for the following user story? Please specify the configuration and make sure the code implements an error-handling framework:
	User story – automate account review notifications
	• **Goal**: As a Sales Manager, I want to automatically send email notifications to Account Owners for accounts that have not been updated in the last 30 days
	• **Acceptance criteria**:
	· **Given** An account has not been updated in the last 30 days
	· **When** the nightly batch job runs
	· **Then** the account owner should receive an email notification reminding them to review and update the account details
Response	Certainly! For the given user story, you'll need a Batch Apex class to identify the accounts that have not been updated in the last 30 days, and then send an email to the account owners. Additionally, you'll need to schedule this Batch Apex to run nightly. Let's dive into the code and configuration details:

Apex Batch class:

```
public class AccountReviewNotificationBatch implements
Database.Batchable<sObject>, Database.Stateful {

    public Database.QueryLocator start(Database.
BatchableContext bc) {
        String query = 'SELECT Id, Name, Owner.Email
FROM Account WHERE LastModifiedDate <= LAST_N_DAYS:30';
        return Database.getQueryLocator(query);
    }

    public void execute(Database.BatchableContext bc,
List<sObject> scope) {
        List<Messaging.SingleEmailMessage> emails = new
List<Messaging.SingleEmailMessage>();

        for (sObject record : scope) {
            Account acc = (Account)record;
            Messaging.SingleEmailMessage email = new
Messaging.SingleEmailMessage();
```

```
            email.setToAddresses(new String[] {acc.
Owner.Email});
            email.setSubject('Reminder: Account Review
Needed for ' + acc.Name);
            email.setPlainTextBody('Please review and
update the account details for ' + acc.Name);

            emails.add(email);
        }

        if (!emails.isEmpty()) {
            try {
                Messaging.sendEmail(emails);
            } catch (Exception e) {
                System.debug('Error sending emails: ' +
e.getMessage());
            }
        }
    }

    public void finish(Database.BatchableContext bc) {
        // You can add any post-processing code here if
required.
    }
}
```

Scheduling the Batch class:

You'll also need to schedule this batch class to run nightly. You can do this programmatically or via Salesforce's UI.

Programmatic scheduling:

System.schedule('Nightly Account Review Notification', '0 0 0 * * ?', new ScheduledClass());

Error handling framework: The preceding batch class includes a simple error handling mechanism that captures any exceptions during the email-sending process and writes them to the debug log. For more advanced scenarios, you can implement a custom logging system where you write these errors to a custom object or send notifications to an admin.

	Configuration:
	Deploy the AccountReviewNotificationBatch class to your Salesforce environment
	Schedule the batch class to run nightly, either programmatically or through the Salesforce UI
	With these steps, you should be able to meet the acceptance criteria specified in the user story.

Great response. This time, we asked for additional technical requirements, such as an error handling framework – which has been implemented very lightly – and a listing of the configuration steps to follow. Again, this is ideal to use as the blueprint of your Apex development.

Let's take it to the next level:

Prompt	Could you please rewrite the previous code using the ExceptionForce error handling framework?
Response	Certainly! Using an error-handling framework such as ExceptionForce can help you manage exceptions in a more structured way, making it easier to debug and monitor your code.
	Assuming that ExceptionForce has methods such as logException to log errors, the updated Batch Apex class would look like this:
| | ```
public class AccountReviewNotificationBatch implements
Database.Batchable<sObject>, Database.Stateful {

    public Database.QueryLocator start(Database.
BatchableContext bc) {
        String query = 'SELECT Id, Name, Owner.Email FROM
Account WHERE LastModifiedDate <= LAST_N_DAYS:30';
        return Database.getQueryLocator(query);
    }

    public void execute(Database.BatchableContext bc,
List<sObject> scope) {
        List<Messaging.SingleEmailMessage> emails = new
List<Messaging.SingleEmailMessage>();

        for (sObject record : scope) {
            Account acc = (Account)record;
            Messaging.SingleEmailMessage email = new
Messaging.SingleEmailMessage();
``` |
```

```
            email.setToAddresses(new String[] {acc.Owner.
Email});
            email.setSubject('Reminder: Account Review
Needed for ' + acc.Name);
            email.setPlainTextBody('Please review
and update the account details for ' + acc.
Name);           emails.add(email);
        }

    if (!emails.isEmpty()) {
        try {
            Messaging.sendEmail(emails);
        } catch (Exception e) {
            ExceptionForce.logException(e,
'Failure in sending email notifications in
AccountReviewNotificationBatch');
        }
    }
}

    public void finish(Database.BatchableContext bc) {
        // You can add any post-processing code here if
required.
    }
}
```

This version of the code uses ExceptionForce.logException to log any exceptions that occur during the process of sending emails. This method will log the exception details and possibly send alerts to administrators, depending on how ExceptionForce is configured.

You'll still need to schedule the batch class as mentioned in the previous example, and make sure the ExceptionForce framework is available in your Salesforce environment.

It didn't do too well this time. It provided a more complex error-handling framework but it didn't put it in the context of a well-known framework. Having said that, this has saved us a lot of time in terms of becoming familiar with these concepts.

Now, let's ask for the test class and identify the risks associated with hitting governor limits:

Prompt	Could you please identify the risks associated with hitting the governor limits of the following Apex code and create the test class?

```
Database.Batchable<sObject>, Database.Stateful {

    public Database.QueryLocator start(Database.
BatchableContext bc) {
        String query = 'SELECT Id, Name, Owner.Email FROM
Account WHERE LastModifiedDate <= LAST_N_DAYS:30';
        return Database.getQueryLocator(query);
    }


    public void execute(Database.BatchableContext bc,
List<sObject> scope) {
        List<Messaging.SingleEmailMessage> emails = new
List<Messaging.SingleEmailMessage>();

        for (sObject record : scope) {
            Account acc = (Account)record;
            Messaging.SingleEmailMessage email = new
Messaging.SingleEmailMessage();

            email.setToAddresses(new String[] {acc.Owner.
Email});
            email.setSubject('Reminder: Account Review
Needed for ' + acc.Name);
            email.setPlainTextBody('Please review and
update the account details for ' + acc.Name);

            emails.add(email);
        }

        if (!emails.isEmpty()) {
            try {
                Messaging.sendEmail(emails);
            } catch (Exception e) {
                ExceptionForce.logException(e,
'Failure in sending email notifications in
AccountReviewNotificationBatch');
            }
        }
    }
```

| | ```
            public void finish(Database.BatchableContext bc) {
                // You can add any post-processing code here if
        required.
            }
        }
``` |
| **Response** | The provided Apex code for a batch process is designed to send out emails to account owners if their accounts have not been modified in the last 30 days. While the code appears relatively simple, there are potential risks regarding governor limits that should be considered:

• **SOQL query limits**: The start() method uses a SOQL query with Database. getQueryLocator. However, it's worth noting that the SOQL query will only count as 1 against the governor limit, no matter how many records are returned, because it's in a batch class.

• **Email limit**: Salesforce has an outbound email limit, both for single and mass emails, when using the Messaging.sendEmail method. Make sure you don't exceed these limits. Given that you're sending an email per account record, this could be a point of concern.

• **CPU time limits**: Although unlikely with this simple script, keep an eye on the CPU time if you plan to add more complex logic in the execute() method.

• **Error logging**: The code attempts to catch any exceptions thrown during the email-sending operation. Make sure the ExceptionForce.logException method doesn't perform operations that could hit other governor limits, such as DML operations. |
```

Here's what a test class for the preceding code could look like:

```
@isTest
private class TestAccountReviewNotificationBatch {
    @isTest static void testBatch() {
        // Create test Account
        Account acc = new Account(Name = 'Test Account',
LastModifiedDate = System.now().addDays(-31));
        insert acc;

        Test.startTest();

        // Instantiate the batch class & execute
        AccountReviewNotificationBatch batchJob = new
AccountReviewNotificationBatch();
        Id batchId = Database.executeBatch(batchJob);

        Test.stopTest();

        // Perform assertions, for example by querying a
logging object or inspecting any emails sent.
        // Note that we cannot check the Messaging.
sendEmail in a test method. It's considered a no-op (no
operation).
    }
}
```

Please make sure you replace AccountReviewNotificationBatch with the actual name of your Batch Apex class.

Two useful features of ChatGPT are its ability to effortlessly generate test classes and its skill in identifying risks associated with the code. These capabilities not only aid in documentation but also in ongoing monitoring, thereby contributing to the overall health of our organization.

Optimizing and refactoring Apex code

During many Salesforce projects, there is a metric that is usually the most difficult to refactor: cyclomatic complexity. Cyclomatic complexity is a software metric that's used to indicate the complexity of a program. It's calculated by developing a control flow graph of the code and then determining the number of paths through the graph. Higher values indicate more complex and potentially more error-prone code.

Let's have a look at a class with high cyclomatic complexity:

```
public class HighComplexityClass {
    public void complexMethod(Integer val) {

        if (val == 1) {
            System.debug('Value is 1');
        } else if (val == 2) {
            System.debug('Value is 2');
            if (val < 3) {
                System.debug('Value is less than 3');
            }
        } else if (val == 3) {
            System.debug('Value is 3');
        } else if (val == 4) {
            System.debug('Value is 4');
            for (Integer i = 0; i < 5; i++) {
                if (i == 2) {
                    System.debug('Loop value is 2');
                } else {
                    System.debug('Loop value is not 2');
                }
            }
        } else {
            System.debug('Value is not between 1 and 4');
        }

        switch on val {
            when 5 {
                System.debug('Value is 5');
            }
            when 6 {
                System.debug('Value is 6');
            }
            when 7 {
                System.debug('Value is 7');
            }
            when else {
                System.debug('Value is not between 5 and 7');
            }
        }
    }
}
```

This code has numerous branches, which would increase its cyclomatic complexity. However, please remember that high cyclomatic complexity is generally considered a bad practice because it makes the code harder to read, test, and maintain. ChatGPT can help refactor the class to reduce this metric:

Prompt	Could you please refactor this class to reduce the cyclomatic complexity?
	<pre>public class HighComplexityClass {
    public void complexMethod(Integer val) {

        if (val == 1) {
            System.debug('Value is 1');
        } else if (val == 2) {
            System.debug('Value is 2');
            if (val < 3) {
                System.debug('Value is less than 3');
            }
        } else if (val == 3) {
            System.debug('Value is 3');
        } else if (val == 4) {
            System.debug('Value is 4');
            for (Integer i = 0; i < 5; i++) {
                if (i == 2) {
                    System.debug('Loop value is 2');
                } else {
                    System.debug('Loop value is not 2');
                }
            }
        } else {
            System.debug('Value is not between 1 and 4');
        }

        switch on val {
            when 5 {
                System.debug('Value is 5');
            }
            when 6 {
                System.debug('Value is 6');
            }
            when 7 {
                System.debug('Value is 7');
            }</pre> |

| | ```
        when else {
                System.debug('Value is not between 5 and
7');
            }
        }
    }
}
``` |
| **Response** | Certainly! One way to reduce the cyclomatic complexity of the class is to break down the logic into smaller methods and minimize the number of conditional branches. Here's a refactored version:<br><br>```
public class RefactoredClass {

    public void complexMethod(Integer val) {
        handleDebugForValuesOneToFour(val);
        handleDebugForValuesFiveToSeven(val);
    }

    private void handleDebugForValuesOneToFour
(Integer val) {
        switch on val {
            when 1 {
                System.debug('Value is 1');
            }
            when 2 {
                System.debug('Value is 2');
                debugForValueTwo(val);
            }
            when 3 {
                System.debug('Value is 3');
            }
            when 4 {
                System.debug('Value is 4');
                debugForLoopValues();
            }
            when else {
                System.debug('Value is not between 1 and
4');
            }
        }
    }
``` |

```
    private void debugForValueTwo(Integer val) {
        if (val < 3) {
            System.debug('Value is less than 3');
        }
    }

    private void debugForLoopValues() {
        for (Integer i = 0; i < 5; i++) {
            debugForSpecificLoopValue(i);
        }
    }

    private void debugForSpecificLoopValue(Integer i) {
        if (i == 2) {
            System.debug('Loop value is 2');
        } else {
            System.debug('Loop value is not 2');
        }
    }

    private void handleDebugForValuesFiveToSeven
(Integer val) {
        switch on val {
            when 5 {
                System.debug('Value is 5');
            }
            when 6 {
                System.debug('Value is 6');
            }
            when 7 {
                System.debug('Value is 7');
            }
            when else {
                System.debug('Value is not between 5 and
7');
            }
        }
    }
}
```

By segregating the logic into smaller methods, we've reduced the cyclomatic complexity and made the code more readable. This also makes it easier to test and maintain.

ChatGPT assists in refactoring code to address concerns such as cyclomatic complexity. By leveraging ChatGPT both during initial development and routine maintenance, we can ensure our code remains clear and comprehensible and serves as a guidepost for best practices in subsequent projects.

Now that we know how ChatGPT can help us create our code, let's have a look at how it can help debug Salesforce code.

What about other PMD violations?

Prompt	Can you check what PMD violations the following code has?
| | ```
public class HighComplexityClass {
    public void complexMethod(Integer val) {

        if (val == 1) {
            System.debug('Value is 1');
        } else if (val == 2) {
            System.debug('Value is 2');
            if (val < 3) {
                System.debug('Value is less than 3');
            }
        } else if (val == 3) {
            System.debug('Value is 3');
        } else if (val == 4) {
            System.debug('Value is 4');
            for (Integer i = 0; i < 5; i++) {
                if (i == 2) {
                    System.debug('Loop value is 2');
                } else {
                    System.debug('Loop value is not 2');
                }
            }
        } else {
            System.debug('Value is not between 1 and 4');
        }
``` |
```

|  | ```
              switch on val {
                  when 5 {
                      System.debug('Value is 5');
                  }
                  when 6 {
                      System.debug('Value is 6');
                  }
                  when 7 {
                      System.debug('Value is 7');
                  }
                  when else {
                      System.debug('Value is not between 5 and
      7');
                  }
              }
          }
      }
``` |
| **Response** | While I can't run PMD, I can provide a manual analysis of the code snippet for typical PMD rule violations. Here's what might be flagged:

Cyclomatic complexity: complexMethod has a lot of conditional branches, which could increase the cyclomatic complexity of the method. PMD might flag it as a violation if it exceeds a certain threshold.

Excessive method length: The method is quite long and does a lot of things, which might be flagged for breaking the single responsibility principle.

Avoid deeply nested if statements: Although not deeply nested, the multiple if-else and inner if statements can also contribute to complexity and may be flagged.

Excessive debug logging: Although not strictly a PMD rule, using excessive debug logging can make it harder to identify real issues during troubleshooting and can fill up your log files quickly. However, if this is just some example code, then this point might not be that relevant.

Lack of comments: PMD might flag the lack of comments explaining what the method does, although this is more of a best practice than a PMD violation.

Magic numbers: The integers used in the conditions might be considered "magic numbers." It's a good practice to replace them with named constants to make the code more readable and maintainable.

Please note that these are manual checks and may not capture all PMD rule violations. |
```

Having explored how ChatGPT can optimize Apex code to adhere to best practices and enhance performance, let's transition to error handling and debugging. The ability to write efficient code is complemented by the skill to troubleshoot and resolve issues when they arise. ChatGPT can be an indispensable tool for quickly identifying bugs and offering solutions to resolve them, ensuring that the Apex code not only runs smoothly but also maintains the robustness required for Salesforce's dynamic environment.

Error handling and debugging in Apex

In a similar way to the previous section, ChatGPT can assist in several ways when it comes to debugging Salesforce code. While the AI isn't a direct replacement for hands-on debugging tools or a development environment, it offers strategic advice and guidance. Here are two important points where ChatGPT can be beneficial:

- **Error explanation**: If you have an error message or exception, you can provide it to ChatGPT, and it can help interpret its meaning and suggest potential fixes based on common causes

- **Logic and algorithm assistance**: If you're trying to achieve a specific function or logic and are getting unexpected results, describing the problem to ChatGPT can help you get suggestions on how to approach or refine your logic

However, while ChatGPT is a valuable tool in the debugging process, it's essential to pair its guidance with hands-on testing, Salesforce developer tools, and sandboxes. The combination of AI insights and practical tools will yield the most effective debugging strategy.

Let's have a look at this by covering a practical example. The execution of a new class is getting the following exception:

```
System.SObjectException: SObject row was retrieved via SOQL without
querying the requested field: Service_Contract_Products__c.Status__c
```

ChatGPT can help understand the problem and fix it:

Prompt	Could you help me understand why I am getting the following exception while executing my class?
	System.SObjectException: SObject row was retrieved via SOQL without querying the requested field: Service_Contract_Products__c.Status__c

Response	Certainly! The exception message you're seeing, "System.SObjectException: SObject row was retrieved via SOQL without querying the requested field: Service_Contract_Products__c.Status__c," provides a clear indication of what went wrong.
	Here's the breakdown:
	What happened? You attempted to access a field on SObject (in this case, the Status__c field on the Service_Contract_Products__c object) after performing a SOQL query.
	Why did the error occur? The Status__c field wasn't included in your SOQL query's SELECT clause. However, you tried to reference or use that field later in your code.
	### How to resolve it?
	Review the SOQL query: Go back to your class and find where you query Service_Contract_Products__c. Ensure that Status__c is part of the fields you're selecting.
	SELECT Id, Name, Status__c FROM Service_Contract_Products__c WHERE ...
	Confirm access: Also, ensure that the user/profile executing the code has read permissions for the Status__c field. If the user doesn't have permission, the field won't be accessible, even if it's present in the SOQL.
	Best practice: As a best practice, always ensure you're querying all fields from SObject that you plan to reference later in your Apex code. This will help prevent such exceptions.
	Lastly, be cautious about adding too many fields to the query if you don't need them. Over-querying can make your code less efficient, especially if you're working with objects that have many custom fields. Always aim for a balance and query only the data you need.

ChatGPT not only identifies the root cause of the problem but also offers best practices to ensure more efficient future development. Additionally, it provides guidance on crafting more effective queries to achieve desired outcomes.

Best practices and continuous learning

ChatGPT can serve as a great aid for Salesforce developers, especially when it comes to following Apex's best practices and fostering a culture of continuous learning. As a conversational AI with the ability to analyze and generate code snippets, ChatGPT can help you understand how to structure your Apex classes, triggers, and batch jobs in a way that adheres to Salesforce's recommended guidelines. From explaining governor limits to demonstrating proper error-handling techniques, ChatGPT can provide real-time insights and suggestions that could save you hours of debugging and troubleshooting.

Additionally, the ever-evolving landscape of software development demands continuous learning; ChatGPT can assist here. Whether you're a beginner looking to understand the basics of Apex or an experienced developer seeking advanced optimization techniques, ChatGPT can tailor its responses to meet your educational needs. With its ability to parse and respond to complex queries, you can explore topics such as test-driven development, bulk processing, and API integrations in a conversational manner. This makes the learning process more interactive and less daunting.

By leveraging ChatGPT, you not only get immediate answers to your Apex-related questions but also expose yourself to broader development practices that you might not have encountered yet. This facilitates a learning environment that is both adaptive and comprehensive, helping you stay up-to-date with the latest Apex practices and, more broadly, become a more proficient Salesforce developer.

For instance, a good example is to find out whether ChatGPT can provide responses that include the last available features of Salesforce.

Prompt	Could you please write an Apex class that queries all fields from the Account object while making sure field- and object-level permissions are enforced?

Response	Certainly! Here's an example of an Apex class that queries all fields from the Account object, while also enforcing both object- and field-level security:
	
	 ```public class AccountQueryWithSecurityChecks {```
	
	``` public static List<Account>```
	```getAccountsEnforcingSecurity() {```
	``` // Describe the Account object```
	``` Schema.DescribeSObjectResult objDescribe = Schema.```
	```SObjectType.Account;```
	
	``` // Check for object-level permission (readable)```
	``` if(!objDescribe.isAccessible()) {```
	``` System.debug('User does not have access to```
	```read Account records.');```
	``` return new List<Account>();```
	``` }```
	
	``` // Create a list to hold the names of accessible```
	```fields```
	``` List<String> accessibleFields = new```
	```List<String>();```

```
        // Get the field map
        Map<String, Schema.SObjectField> fieldMap =
objDescribe.fields.getMap();
        // Iterate over the field map to find accessible
fields
        for(String fieldName : fieldMap.keySet()) {
            Schema.DescribeFieldResult fieldDescribe =
fieldMap.get(fieldName).getDescribe();

            // Check for field-level security (readable)
            if(fieldDescribe.isAccessible()) {
                accessibleFields.add(fieldName);
            }
        }

        // If no fields are accessible, return an empty
list
        if(accessibleFields.isEmpty()) {
            System.debug('User does not have access to
any Account fields.');
            return new List<Account>();
        }

        // Create a SOQL query dynamically
        String query = 'SELECT ' + String.
join(accessibleFields, ', ') + ' FROM Account';

        // Execute the query
        List<Account> accounts = Database.query(query);

        return accounts;
    }
}
```

To use this Apex class, call the getAccountsEnforcingSecurity() method, like so:

```
List<Account> accounts = AccountQueryWithSecurityChecks.
getAccountsEnforcingSecurity();
```

This method will return a list of account records containing only the fields to which the current user has read access. If the user doesn't have read access to the Account object or any of its fields, it will return an empty list and write a debug statement.

Let's be more specific:

Prompt	Could you please rewrite the code using WITH SECURITY_ENFORCED clause?

Response	Certainly! The WITH SECURITY_ENFORCED clause is a straightforward way to enforce field- and object-level security permissions in SOQL queries. Here's how you could update the previous Apex class so that it uses this clause:
	
	``
	public class AccountQueryWithSecurityEnforced {
	
	    public static List<Account>
	getAccountsEnforcingSecurity() {
	        List<Account> accounts;
	
	        try {
	            accounts = [SELECT * FROM Account WITH
	SECURITY_ENFORCED];
	        } catch (System.QueryException e) {
	            // Handle the exception when the user
	doesn't have the necessary permissions
	            System.debug('You don\'t have access
	to one or more fields or the object itself: ' +
	e.getMessage());
	            accounts = new List<Account>();
	        }
	
	        return accounts;
	    }
	}
	``
	
	With this code, the WITH SECURITY_ENFORCED clause will automatically enforce object-level and field-level security. If the current user doesn't have access to either the Account object or any field in the query, Salesforce will throw a System.QueryException error, which can be caught and handled appropriately.
	
	To use this class, simply call the getAccountsEnforcingSecurity() method:
	
	``
	List<Account> accounts =
	AccountQueryWithSecurityEnforced.
	getAccountsEnforcingSecurity();
	``
	
	This method will return a list of account records containing all accessible fields for the current user. If any field or object is not accessible, it will catch the exception and return an empty list.

There you go! ChatGPT is aware of the Salesforce functionalities available in Apex but requires specific information to make sure that your blueprints have them available.

Summary

This chapter covered how to apply ChatGPT to Salesforce Apex development. It provided a foundational grasp of Apex, highlighting ChatGPT's role as an assistant that enhances the coding journey. It covered how ChatGPT can aid in writing Apex code by offering suggestions and ensuring adherence to best practices. This chapter touched on the strategies for optimizing and refactoring Apex code with ChatGPT's input, aiming for improved performance and code sustainability. The *Error handling and debugging in Apex* section detailed how ChatGPT could expedite the troubleshooting process, helping developers pinpoint and solve coding issues effectively. Finally, this chapter reinforced the importance of continuous learning, with ChatGPT serving as a tool for keeping abreast of the latest Apex development practices.

The next chapter will address Salesforce's interactions with various platforms and systems through web services and callouts, illustrating how ChatGPT can be leveraged to streamline this process. It will establish a foundational understanding of web services and callouts, outlining their pivotal role in Salesforce's ability to communicate externally. Following this, we will review how to configure web services within Salesforce, with ChatGPT assisting with best practices and implementation techniques. We will also cover how to use ChatGPT to optimize and manage callouts for better performance and reliability. The next chapter will tackle common issues encountered with web services and callouts, offering strategies for using ChatGPT to quickly diagnose and solve these problems.

7

Using ChatGPT for Salesforce Web Services and Callouts

Salesforce Web Services and Callouts form a vital component of the Salesforce ecosystem, enabling seamless integrations and extended functionalities that go beyond the standard Salesforce capabilities. In an increasingly connected world, where different systems need to talk to each other, these tools are more critical than ever. Let's delve into what they are and how they work.

In this chapter, you will learn the essentials of enabling and refining the communication channels between Salesforce and external systems. Throughout these pages, you'll gain a solid grasp of the foundational elements of Web Services and Callouts, their significance in the Salesforce ecosystem, and the step-by-step processes to set them up effectively. With a focus on practicality, this chapter provides a pathway to optimize these integrations for reliability and performance to help you troubleshoot common issues efficiently using ChatGPT. Through real-world scenarios and case studies, you'll learn to translate user stories into secure, robust integrations, ensuring your Salesforce environment communicates seamlessly with other platforms while adhering to the highest standards of data security and integrity.

Technical requirements

When embarking on the journey of writing Salesforce Apex, the choice of development tools and environment is critical to productivity and success. Apex, Salesforce's proprietary programming language, is designed to execute on Lightning Platform and requires a robust set of tools to manage its life cycle, from development to deployment.

At the core of the Apex development toolkit is the Salesforce **integrated development environment (IDE)**. The most prominent and widely used IDE is **Visual Studio Code** (**VS Code**) with the Salesforce Extension Pack. This pack, officially endorsed by Salesforce, equips developers with features such as syntax highlighting, auto-completion, integrated testing, and debugging capabilities. It also provides direct deployment to Salesforce orgs and version control integration, which are indispensable for maintaining code quality and collaboration. Developers can leverage VS Code's powerful source code editor and customization options to tailor their environment to specific project needs.

Another essential aspect of the development environment is the use of Salesforce sandboxes. These are isolated copies of the Salesforce environment where developers can build and test their Apex code without affecting the live production environment. The type of sandbox that's used can vary from Developer Sandboxes, which are suitable for coding and testing in isolation, to Full Copy Sandboxes, which include a complete copy of the production environment's data for performance and user acceptance testing. Utilizing sandboxes is a best practice as it allows for the safe development and testing of new features, bug fixes, and improvements. The code that's written and tested in these sandboxes can then be pushed to production using change sets or Salesforce DX, depending on the complexity and governance requirements of the project.

IntelliJ IDEA with the Illuminated Cloud plugin is another alternative IDE for Salesforce development. This powerful combination provides a rich set of features tailored for Salesforce development, including smart completions, code navigation, an integrated SOQL query builder, and more, though it tends to be preferred by developers who have an existing affinity for JetBrains' suite of IDEs. Whether opting for VS Code or IntelliJ, the integration of these IDEs with version control systems such as Git is paramount for collaborative and scalable development practices. The toolset, which is complemented by the metadata-aware capabilities of these IDEs, ensures developers can effectively manage their source, track changes, and handle complex deployment processes, thereby facilitating a more streamlined and efficient development life cycle for Salesforce Apex.

Salesforce Web Services

Salesforce Web Services expose custom logic via standard SOAP or REST protocols. They permit external systems to perform operations such as query, search, and CRUD operations directly on Salesforce data. These web services are defined in Apex code and can be customized to offer targeted functionalities. For instance, you might have a web service that triggers specific workflows or updates records based on complex business logic. The beauty of Salesforce Web Services lies in their ability to make the Salesforce platform extendable and open to various kinds of external interactions.

Web services in Salesforce enable different applications to communicate with each other over the web. They play a critical role in the interoperability and integration between Salesforce and other systems. Here are some of the core concepts:

- **Simple Object Access Protocol (SOAP)**: A protocol for exchanging structured information in the implementation of web services. SOAP web services use XML messages and are strongly typed.

- **Representational State Transfer (REST)**: A style of software architecture that uses standard HTTP methods. REST web services use JSON or XML messages and are loosely typed.

- **Apex web service methods**: In Salesforce, you can expose Apex classes and methods so that they can be consumed as web services. This usually involves using the `global` access modifier and annotations such as `@WebService` for SOAP or `@RestResource(urlMapping='/yourUrl')` for REST.

- **Web Service Description Language** (**WSDL**): If you're consuming or exposing a SOAP web service, you'll encounter a WSDL file, which is an XML document that describes how the web service can be called, what parameters it expects, and what data structures it returns.

- **API endpoints**: The URL at which a particular web service can be accessed is known as the API endpoint. For RESTful web services in Salesforce, this would typically be in the form of `https://yourInstance.salesforce.com/services/apexrest/yourUrl`.

- **Authentication and authorization**: Security is often a crucial part of any web service. Salesforce provides various ways to authenticate web service calls, such as OAuth 2.0, Session ID, and basic authentication. This ensures that only authorized users can access or manipulate data.

- **Governor limits**: Salesforce enforces several governor limits on resources such as memory, CPU time, and the number of records retrieved or modified. These limits apply to web services as well, so it's essential to write efficient code.

- **Exception handling**: A robust web service should be able to handle exceptions gracefully and return meaningful error messages to the client. Salesforce allows for custom exception types and fault contracts to handle errors effectively.

- **Asynchronous operations**: Salesforce supports asynchronous operations via `@future` methods, allowing time-consuming operations to be handled in the background, thereby making your web services more scalable and user-friendly.

Understanding these core concepts is crucial for both consuming external web services in Salesforce and exposing Salesforce data and operations as web services to be consumed by external systems.

Salesforce Apex callouts

On the flip side, Salesforce Apex Callouts are essentially outbound requests to external services. They enable Salesforce to communicate with external APIs, be it RESTful or SOAP-based. Callouts are pivotal when Salesforce needs to fetch data from an external service or when an operation in Salesforce should trigger an action in another system. For example, when a new lead is added to Salesforce, a callout could automatically invoke an external email marketing tool to send a welcome email.

Understanding both Salesforce Web Services and Callouts is crucial for anyone who aims to create a deeply integrated, automated, and efficient system. Whether syncing Salesforce with an ERP system, streaming real-time analytics data, or automating cross-system workflows, these tools offer the flexibility and power you need.

Similar concepts from Web Services can be applied to Salesforce Callouts. A callout in Salesforce refers to the act of making a web service request to an external server from your Salesforce org. This is usually done to retrieve or send data from Salesforce to another platform or to integrate Salesforce with other systems. Here are some of the core concepts related to callouts in Salesforce:

- **HTTP methods**: Salesforce supports various HTTP methods, such as GET, POST, PUT, PATCH, and DELETE, for making callouts. The choice of method depends on what operation you're looking to perform on the external system.

- **Named credentials**: Named credentials are a safer and more convenient way to store the endpoint and authentication details needed for callouts. It abstracts the endpoint and simplifies the callout procedure, especially when it comes to maintaining the code.

- **Apex HTTP classes**: Salesforce provides Apex classes such as `HttpRequest`, `HttpResponse`, and `Http` to create and execute HTTP callouts. These classes allow you to set headers, specify the HTTP method, and handle the response.

- **Authentication**: When integrating with external services, you'll often need to authenticate your requests. Salesforce supports various forms of authentication such as Basic Auth, OAuth, or API keys. These can usually be set in the HTTP header or the URL.

- **Asynchronous callouts**: Salesforce allows you to make asynchronous callouts using the `@future` annotation or the `Queueable` interface. Asynchronous callouts are useful for operations that might take a long time to complete and you don't want to hold up the user interface.

- **SOAP and REST callouts**: You can make both SOAP and REST callouts from Salesforce. SOAP callouts use XML-based messaging and typically use a WSDL file for structure. REST callouts are more flexible and usually use JSON or XML for the message format.

- **Governor limits**: Salesforce enforces governor limits on the number and duration of callouts. For synchronous transactions, you can make up to 100 callouts, and each callout can take a maximum of 120 seconds. Asynchronous callouts have different limits.

- **Error handling**: Effective error handling is crucial for a smooth user experience. Salesforce allows you to catch and handle exceptions that occur during a callout, enabling you to return user-friendly error messages and implement retry logic if necessary.

- **Response parsing**: Once you receive a response from an external server, you'll often need to parse this response to extract useful information. Salesforce provides built-in support for parsing both JSON and XML responses.

Understanding these core concepts can significantly improve the efficiency and reliability of making callouts in Salesforce.

ChatGPT can offer significant assistance to Salesforce developers working with Web Services and Callouts in multiple ways. Let's take a look.

Code snippets and prototypes

Developing Web Services or Callouts involves writing Apex code, crafting HTTP requests, or working with SOAP/REST protocols. ChatGPT can generate code snippets and prototype examples for specific scenarios. This helps developers understand how to set up a web service or perform a callout, saving time that might be spent on research or initial code writing.

Debugging assistance

Issues often arise when developing or maintaining Web Services and Callouts – be it poorly formed requests/responses, authorization issues, or logic errors. ChatGPT can help diagnose common problems and offer debugging advice. While it can't directly access your Salesforce instance to debug for you, it can guide you through the thought process and recommend potential solutions based on the symptoms you describe.

Best practices

Salesforce has guidelines and best practices for implementing Web Services and Callouts efficiently and securely. ChatGPT can provide insights into these best practices, such as handling governor limits, ensuring fault tolerance, or managing API versioning, helping you to build more robust and scalable solutions.

Review and refactoring suggestions

Suppose you've already written some code for a web service or a callout. In that case, ChatGPT can help review that code logically and suggest refactorings to improve performance, maintainability, or readability. This could include things such as optimizing SOQL queries, handling exceptions more gracefully, or modularizing code for better reusability.

Interactive learning

ChatGPT can simulate real-world scenarios involving web services and callouts. This enables less-experienced developers to understand the life cycle of these services, including request and response handling, data parsing, and error management, through interactive dialogue.

By assisting in these areas, ChatGPT becomes a valuable companion for Salesforce developers, enabling them to write better, debug faster, and understand deeper. It can be especially useful for solo developers or smaller teams that may not have immediate access to specialized expertise in Salesforce Web Services and Callouts.

Setting up web services with ChatGPT's aid

Consider the following simple example of a user story that allows an external inventory system to update stock levels.

User Story – *integrate Salesforce with an external inventory system*:

- **Goal**: As a backend developer, I want to create a web service in Salesforce so that the external inventory management system can update Salesforce whenever an item's stock level changes.

- **Acceptance criteria**:

  - **Given** the external inventory management system has an update to the stock level of an item

  - **When** the external system sends a POST request with the updated stock level and item ID

  - **Then** the Salesforce web service should update the corresponding record in the Product object with the new stock level

  - **Given** the external system sends an invalid Item ID or stock level

  - **When** the Salesforce web service receives the request

Then the web service should respond with a proper error message and HTTP status code indicating the issue

Given the web service has updated the Salesforce record successfully

When the operation is complete

Then the web service should respond with a success message and HTTP status code indicating the successful update

We can combine the user with a request to ChatGPT to write the code and configuration directions:

Prompt	Could you please write the code and configuration for this user story?
	User story – integrate Salesforce with an external inventory system.
	<ul><li>**Goal**: As a backend developer, I want to create a web service in Salesforce so that the external inventory management system can update Salesforce whenever an item's stock level changes.</li><li>**Acceptance criteria**:<ul><li>**Given** the external inventory management system has an update to the stock level of an item</li><li>**When** the external system sends a POST request with the updated stock level and item ID</li><li>**Then** the Salesforce web service should update the corresponding record in the Product object with the new stock level</li><li>**Given** the external system sends an invalid Item ID or stock level</li><li>**When** the Salesforce web service receives the request</li><li>**Then** the web service should respond with a proper error message and HTTP status code indicating the issue</li></ul></li></ul>

	· **Given** the web service has updated the Salesforce record successfully
	· **When** the operation is complete
	· **Then** the web service should respond with a success message and HTTP status code indicating the successful update
Response	Certainly! Here is an example Apex class and method that would serve as a web service to handle the stock level updates from an external inventory management system. This code is meant to illustrate how the user story could be implemented; it has not been tested in a live Salesforce environment.

Apex class for a web service:

```
@RestResource(urlMapping='/inventory/')
global with sharing class InventoryWebService {
    @HttpPost
    global static String updateStockLevel() {
        RestRequest req = RestContext.request;
        RestResponse res = RestContext.response;

        // Parse JSON payload from request
        Map<String, Object> requestBody = (Map<String,
Object>) JSON.deserializeUntyped(req.requestBody.
toString());

        // Extract the Item ID and new stock level
        String itemId = (String) requestBody.
get('itemId');
        Integer newStockLevel = Integer.
valueOf(requestBody.get('newStockLevel'));

        // Validate Item ID and Stock Level
        if(String.isBlank(itemId) || newStockLevel ==
null || newStockLevel < 0) {
            res.statusCode = 400;
            return 'Bad Request: Invalid Item ID or Stock
Level';
        }

        // Query the Product object to find the matching
record
        List<Product__c> products = [SELECT Id, Stock_
Level__c FROM Product__c WHERE Id = :itemId LIMIT 1];
```

```
                // Check if the Product exists
                if(products.isEmpty()) {
                    res.statusCode = 404;
                    return 'Not Found: Item ID not found';
                }

                // Update Stock Level
                Product__c productToUpdate = products[0];
                productToUpdate.Stock_Level__c = newStockLevel;
                update productToUpdate;

                // Return Success
                res.statusCode = 200;
                return 'Success: Stock Level Updated';
            }
        }
```

Configuration steps:

1. The Product__c object should exist in your Salesforce schema with a field named Stock_Level__c to hold the stock level. This field should be of the Number or Integer type.

2. Make sure the Apex class has the appropriate permissions for the profiles that will interact with this web service.

3. Expose the web service to the external system and share the REST endpoint URL.

4. Ensure that the external system sends POST requests to this endpoint with the JSON payload containing itemId and newStockLevel.

This code and configuration should help you meet the acceptance criteria specified in the user story.

If we take a moment to analyze the response from ChatGPT, we will see that it includes both the code and the configuration steps. It made some good decisions, such as including a few validations and errors related to data consistency, and it didn't pick a Standard Salesforce object such as `Product2` that wouldn't fit this business requirement.

As a minor issue, ChatGPT suggests the `Stock Level` field to be of the `Number` or `Integer` type. While `Number` is OK, `Integer` is not a type of field in Salesforce.

Now, let's ask for the test class and JSON to test if it works:

Prompt	Could you also provide the test class and sample JSON to test it?

Response	Absolutely! Here is an Apex test class to test the web service, followed by a sample JSON payload that could be used to test the web service.
	
	Apex test class:
	
	``apex
	@IsTest
	public class InventoryWebServiceTest {
	
	    @IsTest static void testUpdateStockLevel() {
	        // Create a test Product record
	        Product__c testProduct = new
	Product__c(Name='TestProduct', Stock_Level__c=10);
	        insert testProduct;
	
	        // Prepare the test request and response
	        RestRequest req = new RestRequest();
	        RestResponse res = new RestResponse();
	
	        req.requestURI = '/services/apexrest/
	inventory/';
	        req.httpMethod = 'POST';
	        req.requestBody = Blob.valueOf(JSON.
	serialize(new Map<String, Object>{
	            'itemId' => testProduct.Id,
	            'newStockLevel' => 20
	        }));
	        // Set up test context
	        RestContext.request = req;
	        RestContext.response = res;
	
	        // Execute the web service method
	        String responseMessage = InventoryWebService.
	updateStockLevel();
	
	        // Asserts
	        System.assertEquals('Success: Stock Level
	Updated', responseMessage);
	        System.assertEquals(200, res.statusCode);
	``

```
                // Query the updated record
                Product__c updatedProduct = [SELECT Stock_
        Level__c FROM Product__c WHERE Id = :testProduct.Id
        LIMIT 1];
                System.assertEquals(20, updatedProduct.Stock_
        Level__c);
            }
        }
```

Sample JSON payload

Here's an example JSON payload that you could use to test this web service via tools such as Postman:

```
{
    "itemId": "a00xxxxxxxxxxxx",  // Replace with actual
Product__c record Id
    "newStockLevel": 20
}
```

Simply replace the itemId value with an actual Product__c record ID from your Salesforce instance, and set newStockLevel to the stock level you want to update to.

Remember, the preceding example and test class assume that you have a custom object, Product__c, with a custom field, Stock_Level__c. Make sure you adjust the code so that it fits your actual Salesforce schema.

In this case, the learning approach of ChatGPT even helps with manipulating the sample JSON, which could be very interesting for testers to use if they are not very familiar with the functionality of Salesforce.

When this code and test class are deployed to a developer sandbox, they work. However, the test class only covered 77% of the code of the web service as it is only testing positive results. The authors of this book will leave this to you as an exercise in using ChatGPT to increase test class coverage to 95%.

Optimizing and managing callouts using ChatGPT

Managing callouts effectively in Salesforce is essential for ensuring optimal performance, scalability, and reliability. Here are some best practices for managing callouts:

- **Bulkify your callouts**: When possible, design your callouts so that they handle multiple records at a time. This will help you stay within Salesforce's limits for the number of callouts in a single transaction.

- **Asynchronous processing**: Use asynchronous Apex, such as Future methods or Queueable Apex, to perform callouts. This allows the callout to execute in the background, thereby not blocking the user interface and improving the overall user experience.

- **Error handling**: Implement robust error-handling mechanisms to manage timeouts, status codes, and any other issues that might arise during the callout. Always check the response status and handle exceptions gracefully.

- **Reusable code**: Create reusable utility classes and methods for making callouts. This enables better code maintainability and easier updates.

- **Timeouts**: Be mindful of the callout timeouts. The default is 10 seconds but you can set it up to a maximum of 120 seconds.

- **Callout patterns**: For more complex callouts, consider using design patterns such as Singleton for shared configurations or Facade for simplifying complex callouts.

- **Named Credentials**: Use Named Credentials to securely store the endpoint and authentication settings. This makes it easier to manage sensitive information.

- **Unit testing**: Write comprehensive test methods to cover various scenarios, including error conditions. Use mocking to simulate the behavior of the external system.

- **Monitoring and logging**: Implement logging mechanisms to capture essential information about the callout and its response. This can be useful for debugging and auditing purposes.

- **API limits**: Always keep an eye on the API limits, especially if you are working with external services that have their own limitations.

We are not going to cover all of them in this book; we will only focus on callout patterns and reusable code. But all of them are important to keep in mind to make sure the performance of the organization is not impacted negatively.

Let's start with callout patterns and put ourselves in the shoes of a developer who is unsure of the best pattern to use:

Prompt	Can you help me understand a callout pattern in Salesforce?
Response	Certainly! A callout pattern in Salesforce is essentially a design strategy for managing callouts to external services. Effective patterns for callouts can make your integrations more efficient, maintainable, and reliable. Here, I've outlined a high-performance callout pattern commonly used in Salesforce. **The batch and queueable pattern** **Overview**: This pattern combines batch processing and Queueable Apex to execute callouts in bulk and asynchronously. It's especially useful when you need to make a large number of callouts or process a large volume of data. Here's how it works: • Batch Apex groups records into small chunks • Each chunk of records is passed to Queueable Apex, which performs the callout

- Queueable Apex can chain itself, allowing for multiple callouts to be processed asynchronously in sequence

Implementation steps:

- **Batch class**: Create a Batch Apex class that queries the necessary Salesforce records:

```
public class MyBatchClass implements Database.
Batchable<sObject>{
        public Database.QueryLocator start(Database.
BatchableContext bc){
                return Database.getQueryLocator('SELECT
Id, Name FROM MyObject__c WHERE SomeField__c =
SomeValue');
        }
        public void execute(Database.
BatchableContext bc, List<sObject> records){
                List<Id> recordIds = new List<Id>();
                for(sObject record : records){
                    recordIds.add(record.Id);
                }
                System.enqueueJob(new
MyQueueableClass(recordIds));
        }
        public void finish(Database.BatchableContext
bc){
                // Post-processing code
        }
    }
```

- **Queueable class**: Create a Queueable Apex class that performs the callout:

```
public class MyQueueableClass implements
Queueable {
        private List<Id> recordIds;

        public MyQueueableClass(List<Id> recordIds){
            this.recordIds = recordIds;
        }

        public void execute(QueueableContext
context) {
                // Perform callout logic here
        }
    }
```

	Advantages:
	• Bulk processing: By using Batch Apex, you can process records in bulk, thereby optimizing API calls to external services
	• Asynchronous: The asynchronous nature of Queueable Apex allows callouts to be made in the background
	• Error handling: You can implement robust error handling within the Queueable class, capturing issues and possibly retrying the callout
	Considerations:
	• API limits: Be mindful of the limits both in Salesforce and the external service you are calling
	• Order of execution: Queueable jobs are executed asynchronously and not necessarily in the order they are enqueued
	By following a pattern like this, you can optimize the performance and reliability of your Salesforce callouts.

Combining batch and queueable Apex is not recommended unless it is really necessary. This emphasizes the importance of using ChatGPT as a guide and investigating/validating the decisions made based on ChatGPT's responses.

Note that the response from ChatGPT, which emerged from a prompt lacking in detail, stands out as particularly detailed and comprehensive. The prompt did not specify a use case, nor did it delineate any particular requirements for data synchronization or volume. Nonetheless, ChatGPT selected a callout pattern apt for larger volumes of data – a decision that, upon reflection, aligns precisely with the query's implied needs. To enhance clarity and precision in future inquiries, it would be beneficial to provide more specific details:

Prompt	Can you write a code pattern in Salesforce to manage real-time callouts to an external system with a REST API? We are anticipating a low volume of calls per day, but changing JSON.
Response	Certainly! When you're dealing with real-time callouts to an external REST API with low volumes, and you expect the JSON structure to change frequently, you might want to use a flexible and modular design. Here is a sample code pattern that outlines a way to perform real-time callouts.

Code pattern: |

Apex class for HTTP callout:

```
public class RealTimeCalloutService {
    public static HttpResponse makeCallout(String
endPoint, String httpMethod, String jsonBody) {
        Http http = new Http();
        HttpRequest req = new HttpRequest();
        HttpResponse res = new HttpResponse();

        req.setEndpoint(endPoint);
        req.setMethod(httpMethod);
        req.setHeader('Content-Type', 'application/
json');
        req.setBody(jsonBody);

        try {
            res = http.send(req);
        } catch(Exception e) {
            // Handle callout exception here
            System.debug('Callout error: ' +
e.getMessage());
        }

        return res;
    }
}
```

Utilize custom metadata or custom settings:

Given that the JSON might change frequently, you could store the JSON schema or mappings in custom metadata or custom settings. This allows admins to make changes without altering code.

Dynamic JSON parsing:

To handle the changing JSON structure, use Salesforce's JSON.deserializeUntyped method to parse it dynamically:

```
Map<String, Object> parsedResponse = (Map<String,
Object>) JSON.deserializeUntyped(response.
getBody());
```

Example trigger to invoke a callout:

```
trigger MyRealTimeTrigger on MyObject__c (after
insert, after update) {
    List<String> jsonBodies = new List<String>();

    for (MyObject__c obj : Trigger.new) {
        // Prepare JSON
        String jsonBody = '{ "field1": "' + String.
escapeSingleQuotes(obj.Field1__c) + '", "field2":
"' + String.escapeSingleQuotes(obj.Field2__c) +
'"}';
        jsonBodies.add(jsonBody);
    }

    // Call future method if there are records to
process
    if (!jsonBodies.isEmpty()) {
        CalloutHandler.makeCallouts(jsonBodies);
    }
}
```

Considerations:

Error handling: Add error-handling logic that captures any exceptions or error status codes returned from the callout

Governor limits: Since this is a real-time callout, ensure you are aware of the synchronous callout limits

Schema changes: If the schema is expected to change frequently, consider using custom metadata or custom settings to manage those changes dynamically without requiring code changes

With this pattern, you can manage real-time callouts in a more modular and dynamic way.

The better the ChatGPT prompt, the better the ChatGPT response. ChatGPT has provided a pattern that is ideal for real-time callouts, helps create JSON dynamically, and makes suggestions to use custom metadata or custom settings to help with the dynamic nature of the issue being addressed.

We can go one step further and ask for a framework to manage the callouts:

Prompt	Can you provide a reusable framework for callouts?
Response	Certainly! Creating a reusable framework for callouts can make your Salesforce implementation much more maintainable and flexible. Here is a simplified example of using an interface and classes to handle REST API callouts. This pattern is designed for low-volume, real-time callouts with potentially changing JSON payloads, as per your requirement:

```
// ICalloutService Interface
public interface ICalloutService {
    HttpResponse makeCallout(Map<String, Object> params);
}
// Abstract Callout Service
public abstract class AbstractCalloutService implements
ICalloutService {
    public HttpRequest createRequest(String endpoint,
String method, String body) {
        HttpRequest req = new HttpRequest();
        req.setEndpoint(endpoint);
        req.setMethod(method);
        req.setHeader('Content-Type', 'application/
json;charset=UTF-8');
        if (String.isNotBlank(body)) {
            req.setBody(body);
        }
        return req;
    }
}
// Concrete Callout Service for Inventory Management
public class InventoryCalloutService extends
AbstractCalloutService {
    public HttpResponse makeCallout(Map<String, Object>
params) {
        String endpoint = 'https://api.inventory.com/
items/' + params.get('itemId');
        String body = JSON.serialize(params);

        HttpRequest req = createRequest(endpoint, 'POST',
body);

        Http http = new Http();
        HttpResponse res = http.send(req);

        return res;
```

```
        }
    }
// Handler class to determine which service to use
public class CalloutHandler {
    public static HttpResponse executeCallout(String
serviceName, Map<String, Object> params) {
        ICalloutService service;

        if (serviceName == 'Inventory') {
            service = new InventoryCalloutService();
        }
        // Add more services as needed

        if (service != null) {
            return service.makeCallout(params);
        } else {
            return null;
        }
    }
}


// Sample usage in some Apex code
public class MyController {
    public void updateInventory() {
        Map<String, Object> params = new Map<String,
Object>{
            'itemId' => '123',
            'stockLevel' => 50
        };

        HttpResponse res = CalloutHandler.
executeCallout('Inventory', params);

        // Handle response
    }
}
```

How to use:

- The ICalloutService interface defines the contract for all callout services

- AbstractCalloutService provides a basic HttpRequest creation method

- Concrete classes such as InventoryCalloutService implement the actual callout logic

- The `CalloutHandler` class delegates the callout to the appropriate service based on some condition (here, this is `serviceName`)

- The `MyController` class shows how to use this framework to make a callout

With this pattern, you can easily add more types of callouts by creating new classes that implement `ICalloutService`. This makes it simple to manage changes in JSON payload structure or even completely different types of callouts.

Utilizing Apex frameworks in Salesforce can significantly streamline the development process by providing a structured, reusable code base, which can lead to consistent and maintainable coding practices. Apex frameworks can also enhance efficiency, allowing developers to focus on unique business logic rather than boilerplate code, and facilitate easier debugging and testing due to their standardized architecture. These frameworks often come with built-in best practices, ensuring that solutions are not only robust but also scalable and aligned with the evolving ecosystem of Salesforce.

Troubleshooting web service and callout issues

ChatGPT can serve as a valuable resource for developers seeking to troubleshoot issues related to web services and callouts in Salesforce. Whether you're facing challenges in request formatting, HTTP method selection, header configuration, or response handling, ChatGPT can guide you through common pitfalls and offer suggestions for best practices.

For example, if you encounter an unexpected HTTP status code, ChatGPT can help you understand the likely causes of that status code and ways to resolve the issue. Similarly, if your SOAP or REST callout isn't functioning as expected, ChatGPT can provide insights into common issues such as malformed XML/JSON, unauthorized access, or governor limit exceedances, and suggest ways to debug or refactor your code.

Moreover, as we saw previously, ChatGPT can assist you in writing test classes for your callouts, an often overlooked but crucial part of Salesforce development. By offering code samples or pointing out where mock classes and methods could be implemented, ChatGPT enhances your ability to produce robust, error-free services. The tool is not only reactive, helping you solve issues, but also proactive, equipping you with the knowledge and best practices to avoid future problems.

With its advanced capabilities and extensive knowledge base, ChatGPT aims to be a reliable assistant for troubleshooting your web services and callouts, ultimately aiding you in crafting more robust and efficient Salesforce applications.

Let's start with the example of some malformed JSON:

```
{
    "name": "John",
    "age": 30,
```

```
    "email": john.doe@domain.com  // Invalid because the email is not
enclosed in quotes
  }
```

Here, we are testing our web service by creating our customer, John, but it is not working correctly and the response contains the typical error: malformed JSON. ChatGPT can easily point us in the right direction:

Prompt	Could you please help me understand why the following JSON is malformed and provide me with well-created JSON? ```\n{\n "name": "John",\n "age": 30,\n "email": john.doe@domain.com\n}\n```
Response	Certainly! The issue with the provided JSON is that the email value, john.doe@domain.com, is not enclosed in double quotes, which is required for string values in JSON. Because of this, the JSON parser will not be able to understand the value and it will be considered malformed. Here is the corrected version: ```\n{\n "name": "John",\n "age": 30,\n "email": "john.doe@domain.com"\n}\n``` In this version, all string values and property names are enclosed in double quotes, making it a well-formed JSON object.

It's easy for ChatGPT to come up with a good example and explain the root cause of the problem. But why don't we take it a couple of steps further? One common issue when dealing with callouts is getting the following error:

```
Error - 'You have uncommitted work pending'
```

ChatGPT can help identify the issue and suggest a solution:

Prompt	Why am I getting this error: Error - 'You have uncommitted work pending?

Response	The error "You have uncommitted work pending" often occurs in Salesforce when you are trying to perform a callout after making a **Data Manipulation Language (DML)** operation in the same transaction. Salesforce does not allow you to do any DML operations before a callout in the same transaction to ensure data consistency.
	Here are some typical scenarios where this can happen:
	• You perform a DML operation (insert, update, delete, and so on) and then immediately attempt to make an HTTP callout.
	• You start a database transaction, perform some operation, and then try to make a callout before committing the transaction.
	To fix this issue, you have several options:
	• **Move the callout to a Future method**: You can use a @future method to move the callout to a separate transaction, allowing the DML operation to commit before the callout occurs. Just remember that @future methods have governor limits:

	<pre>@future(callout=true)
	public static void makeCalloutMethod() {
	    // your callout logic
	}</pre>

	• **Use Queueable Apex**: Like @future, Queueable Apex runs in a separate transaction. This allows for more flexibility and is the modern replacement for future methods:

	<pre>public class MyQueueable implements Queueable,
	Database.AllowsCallouts {
	    public void execute(QueueableContext
	context) {
	        // your callout logic
	    }
	}</pre>

	• **Rearrange your logic**: Sometimes, you can rearrange your code logic to perform all DML operations after all the callouts are done.
	• **Perform callouts first**: If possible, perform the callout first and then do the DML operations.
	By ensuring that DML operations and callouts are separated into different transactions, you can avoid this error.

With that, we can summarize what we've learned.

Summary

In this chapter, you enriched your understanding of web services and callouts within the Salesforce ecosystem, learning how to effectively set up, optimize, and troubleshoot these essential integrative features. You have acquired knowledge of the core concepts, delved into configuring web services with accuracy, and explored optimization techniques for improved performance and reliability. The guidance provided by ChatGPT has empowered you to swiftly identify and resolve common integration issues, ensuring your Salesforce solutions communicate seamlessly with other systems while adhering to robust security protocols. You have translated user stories into successful Salesforce integrations, thus enhancing your proficiency and confidence in managing complex digital interactions.

In the next chapter, you will delve into the realm of Salesforce triggers, understanding their pivotal role in automating processes and how ChatGPT can elevate their creation and maintenance. This chapter will guide you through the intricacies of trigger design, drawing on ChatGPT's insights to ensure these triggers are not only effective but also adhere to best practices. You will learn the critical skills of optimizing and refactoring existing triggers for peak performance, and gain valuable troubleshooting techniques to swiftly handle any arising errors. This chapter will impart foundational knowledge, coupled with the capability to design, optimize, and debug triggers, thereby enhancing the Salesforce environment's responsiveness and reliability.

Using ChatGPT for Salesforce Triggers

In the world of Salesforce, `Trigger` stand out as one of the fundamental building blocks, acting as the vanguard of automated process initiation. Simply put, a Salesforce `Trigger` is a piece of code that executes automatically either before or after specific events in Salesforce records, such as inserts, updates, or deletions.

A trigger's primary function is to manage, modify, and streamline the data within the Salesforce platform when the data is created, updated, deleted, or undeleted. Here's how they work: imagine an operation in Salesforce, such as updating a contact's information. Before or after this event takes place, a trigger can intervene, executing predefined logic. For instance, if a contact's city changes, a trigger might automatically update their corresponding mailing address.

The initiation of triggers depends on various record operations. Here are the chief catalysts:

- **Before triggers**: These are used to update or validate the values of a record before they're saved to the database. They're executed before the actual operation (such as `INSERT` or `UPDATE`) occurs. This preemptive approach is particularly useful for ensuring data quality and consistency.

- **After triggers**: Once the data operation is completed and the data has been committed to the database, "after" triggers spring into action. They are typically used to access newly added or modified records and perform operations on them, such as updating related records or interacting with external systems.

- **Events that initiate triggers**: Several events can act as the starting point for triggers, such as the following:

  - `Insert`: When a new record is added

  - `Update`: When a record is modified

  - `Delete`: When a record is removed

- `Undelete`: When a deleted record is restored

- `Merge`: When records are merged

Salesforce triggers act as the sentinels of data operations within the platform. By setting up these automatic responses to changes in data, organizations can ensure that their Salesforce records remain accurate and up-to-date and adhere to specific business logic, thus enhancing overall data integrity and operational efficiency.

In this chapter, you will be equipped with an understanding of Salesforce triggers and how you can enhance your development process using ChatGPT. This chapter will impart knowledge on designing precise and efficient triggers, optimizing and maintaining existing ones, and effectively handling errors and debugging. It emphasizes translating requirements into actionable trigger configurations and focuses on acquiring a disciplined approach to trigger management while incorporating best practices. By the end, you will be adept at not only creating and refining triggers but also at ensuring their governance within a dynamic Salesforce environment.

Exploring business processes that require Salesforce triggers

Salesforce triggers offer a compelling solution for automating and enhancing various business processes. The complexities of the modern business landscape often demand more than the standard functionalities of Salesforce's out-of-the-box features. Here are examples of business scenarios that necessitate the crafting of bespoke triggers to meet specific operational needs:

- **Automated task assignment**: Consider a scenario where a global company utilizes Salesforce to manage its leads. Depending on the geographical location of a lead, the company may wish to automatically assign it to a regional sales representative. An `Insert` trigger can be fashioned to monitor when new leads are added. Upon detection, the trigger would automatically assign the lead to a sales representative based on predefined regional criteria, ensuring the lead is directed to the most relevant person without manual intervention.

- **Maintaining hierarchical data relationships**: Imagine a company that logs major transactions with its clients in Salesforce. Each time a transaction exceeds a certain monetary threshold, it's imperative to notify the client's account manager and update the client's VIP status. An `Update` trigger can be implemented here. When a transaction is logged and crosses the specified monetary limit, this trigger would both update the client's status to "VIP" and send an automated notification to the account manager, thereby ensuring timely acknowledgment and possibly enhanced service for top-tier clients.

- **Data cleansing and standardization**: Data consistency is pivotal for businesses. A company might want all phone numbers stored in Salesforce to follow a standardized format, regardless of how they're input. For this, a `Before Insert` or `Before Update` trigger can be valuable. Whenever a new contact is added or an existing contact is modified, the trigger would reformat the phone number field to the company's chosen standard before the data is committed to the database.

- **Cascading record updates**: In some businesses, the update of one record might necessitate changes in related records. For instance, if a product's price is altered in a catalog, all pending orders with that product might need a price adjustment to reflect the change. An `Update` trigger on the product record could ensure that when a product's price changes, all related order records are automatically updated with the new pricing, safeguarding against potential revenue discrepancies.

In essence, Salesforce triggers play an indispensable role in adapting the CRM platform to the unique, dynamic, and sometimes intricate business processes of various organizations. By providing tailored automated responses to data events, triggers augment operational efficiency, accuracy, and responsiveness.

Salesforce trigger handlers – orchestrating efficacy

In the Salesforce architecture, while triggers stand as the vanguard of data-driven automation, trigger handlers orchestrate this automation. Diving deeper, a clearer picture of their significance in a streamlined Salesforce implementation arises:

1. **Defining trigger handlers**: At a foundational level, a trigger handler is a class – a distinct unit of code – that contains the logic that the trigger will execute. Instead of placing all logic directly within the trigger, the trigger calls methods within these classes to perform the desired actions. By externalizing the logic into a handler, the code becomes more organized, modular, and maintainable. Each function or operation resides in its compartment, and the trigger itself remains lean, serving primarily as a dispatcher to these compartments.

2. **The rationale for using trigger handlers**: Adopting trigger handlers underscores best practices for several compelling reasons:

   - **Maintainability**: As business processes evolve, the underlying logic in Salesforce might need revision. With trigger handlers, modifications can be made directly within a specific method in the handler class, rather than sifting through dense, monolithic trigger code.

   - **Testability**: Salesforce mandates that custom code, including triggers, have a certain percentage covered by unit tests before it's deployed. By isolating logic in handler classes, it's easier to write and manage unit tests for specific functionalities.

   - **Scalability**: As businesses grow, so does the complexity of their operations. Trigger handlers allow for scalable architecture, where new functionalities can be seamlessly added as distinct methods in the handler class without disrupting existing operations.

   - **Order of execution**: Salesforce triggers can fire multiple times in a single transaction, especially in scenarios involving cascading actions. Trigger handlers can help manage and control the order of execution, ensuring that operations happen predictably and efficiently.

3. **Mechanics of trigger handlers**: The working principle behind trigger handlers is one of delegation. When a trigger is fired due to a data operation, it delegates the responsibility of executing the business logic to its associated handler. For instance, on updating a record, a trigger might invoke `updateMethod()` within its handler class. This method contains the logic to execute, be it data validation, record modification, or any other task. Post-execution, control returns to the trigger, which completes its cycle and concludes the operation.

In the grand tableau of Salesforce development, trigger handlers play an instrumental role in ensuring that the automation landscape is not just functional but also elegant and efficient. By compartmentalizing logic and promoting modularity, they champion the principles of sustainable and scalable software design.

Harnessing ChatGPT for Salesforce trigger development

In the rapidly evolving realm of CRM systems, the utilization of sophisticated AI-driven tools such as ChatGPT opens up innovative avenues. Salesforce triggers, though pivotal, can sometimes be complex, demanding tailored logic for specific business processes. Integrating a platform such as ChatGPT into this development cycle offers intriguing potential:

*   **Rapid prototyping**: One of the most compelling uses of ChatGPT in the Salesforce trigger creation process is rapid prototyping. Developers can describe a particular business logic or use case scenario in natural language to ChatGPT, which can then help draft a basic version of the trigger or offer a pseudocode outline. This prototype serves as a foundation, accelerating the development process by providing a blueprint for developers to refine and elaborate upon.

*   **Knowledge augmentation**: ChatGPT, with its vast knowledge base, can assist in clarifying doubts or suggesting best practices related to Salesforce trigger development. Whether it's a query about the trigger execution order, governor limits, or context-specific variables, ChatGPT can act as an on-demand consultant, reducing the need for extensive external research and ensuring adherence to Salesforce's best practices.

*   **Code review assistance**: While ChatGPT isn't a replacement for thorough human-driven code reviews, it can still assist in preliminary examinations. Developers can share snippets of their trigger code with ChatGPT to seek insights and potential optimizations or to identify any blatant red flags. By facilitating these initial checks, ChatGPT can enhance the overall code quality, ensuring triggers are efficient, effective, and free from common pitfalls.

*   **Documentation and explanation**: Often, the logic that's encapsulated within triggers needs to be understood by stakeholders who might not have a deep technical background. ChatGPT can be leveraged to translate complex trigger functionalities into more comprehensible explanations or summaries. These AI-driven descriptions can aid in crafting documentation and user manuals or simply in ensuring that the broader team understands the implemented automation.

*   **Advantages of integrating ChatGPT**: By employing ChatGPT in the Salesforce trigger development process, organizations stand to benefit from quicker development cycles, consistent adherence to best practices, and enhanced knowledge sharing. Moreover, the AI's ability to

comprehend and generate code based on natural language interactions reduces the entry barrier, allowing even those with limited coding expertise to participate in the preliminary stages of trigger development.

As ChatGPT continues to make strides into software development domains, its integration into CRM platforms such as Salesforce is a logical progression. Leveraging ChatGPT for trigger development amalgamates the power of AI with the flexibility of Salesforce, ushering in a new era of efficient, informed, and inclusive CRM automation.

Salesforce trigger best practices

Salesforce triggers, if designed carefully, can significantly streamline and automate business processes within the CRM. However, if not handled carefully, they can lead to unanticipated behaviors, performance issues, and governor limit breaches. Adhering to best practices is vital to ensure reliable, efficient, and scalable triggers. Here's a comprehensive look at these best practices:

- **One trigger per object**: Instead of having multiple triggers for a single Salesforce object, consolidate them into a single trigger. This avoids issues with the order of execution and makes it easier to manage and maintain the code. The trigger can then delegate specific tasks to helper classes or methods based on the operation (insert, update, delete, and so on).

- **Avoid hardcoding IDs**: Hardcoding record IDs (or any other data) is a dangerous practice. These can change between environments (sandbox, production, and so on), leading to deployment issues and runtime errors. Always retrieve dynamic values programmatically or use custom settings or custom metadata for configuration values.

- **Bulkify your code**: Salesforce operates on a bulk processing model. Always ensure your trigger can handle multiple records at once. Avoid putting SOQL queries or DML operations inside loops as this can make you quickly hit governor limits.

- **Use context-specific variables**: Salesforce provides trigger context variables, such as `Trigger. new` and `Trigger.old`, to allow developers to access the records that fired the trigger. Understand and use these variables effectively to cater to different scenarios (such as distinguishing between insert and update operations).

- **Opt for trigger handlers**: As discussed earlier, rather than placing all your logic directly within the trigger, delegate the heavy lifting to trigger handler classes. This approach keeps the trigger clean and promotes modularity and reusability.

- **Implement proper error handling**: Ensure your trigger code gracefully handles exceptions and provides meaningful error messages to the end users. This aids in troubleshooting and ensures users aren't left with cryptic or generic system error messages.

- **Limit SOQL and DML operations**: Be mindful of the governor limits. Always optimize the number of SOQL queries and DML operations. By reducing these, you ensure that your triggers operate efficiently and do not exhaust system resources.

- **Test thoroughly**: Every trigger should have associated test classes that cover various scenarios. Salesforce mandates a code coverage of at least 75%, but aims for higher. Beyond coverage, ensure your tests validate the trigger's actual functionality and edge cases.

- **Avoid recursive triggers**: Recursive triggers, where a trigger inadvertently fires itself, can cause unexpected behaviors and quickly hit governor limits. Implement checks or static variables to prevent unwanted recursive calls.

- **Order of execution awareness**: Understand the Salesforce order of execution (the sequence in which automation tasks such as validation rules, workflows, and triggers are processed). This helps ensure that your trigger behaves as expected, especially when multiple automations are in play.

- **Comment and document**: Always document the purpose and functionality of your triggers (and associated helper classes). This assists other developers in understanding the logic and intent behind the code. Regular comments within the code also help clarify complex sections and decision points.

- **Stay updated and review regularly**: Salesforce undergoes multiple updates annually. Regularly review your triggers to ensure they're optimized for the latest platform capabilities and aren't using deprecated features.

In conclusion, creating triggers in Salesforce isn't merely about scripting automation. It's about crafting a reliable, efficient, and maintainable piece of the broader Salesforce ecosystem. By adhering to the aforementioned best practices, developers can ensure that their triggers not only meet immediate business needs but also stand the test of time, scalability, and platform evolution.

What role does ChatGPT play in this?

Leveraging ChatGPT in the development of Salesforce triggers can greatly expedite the process while maintaining, or even enhancing, the quality of the code. Here's how ChatGPT can be applied:

1. **To automate the code generation process**: Through its deep understanding of Salesforce's Apex programming language, ChatGPT can be utilized to automatically generate code snippets for triggers. Developers can provide specific requirements, such as the object being triggered, the trigger event (for example, before insert, after update, and so on), and the desired action. ChatGPT can then translate these requirements into a well-structured and syntactically correct Apex trigger code. This not only speeds up the development process but also reduces the likelihood of human errors.

2. **To enhance collaboration and understanding**: ChatGPT's natural language processing capabilities enable a bridge between technical and non-technical stakeholders. Translating complex requirements into actionable code and explaining code in human-readable terms fosters better understanding and collaboration among team members. This is particularly beneficial when you're working with complex triggers where business logic and rules must align precisely with organizational needs.

3. **To test and validate assistance**: ChatGPT can aid in creating test classes for the triggers, which is a vital part of Salesforce development to ensure code coverage and functionality. By understanding the trigger's logic, it can generate relevant test scenarios and corresponding Apex test code. This assists developers in validating the triggers quickly, ensuring that they meet the defined criteria and function as expected.

4. **For continuous learning and improvement**: ChatGPT's ability to process vast amounts of information and continuously learn enables it to stay abreast of best practices and Salesforce updates. This means that the code that's generated is not only accurate but also adheres to the latest standards and recommendations. It can even provide insights and suggestions for optimizing existing triggers, leveraging the best patterns, and avoiding common pitfalls.

5. **To facilitate compliance and security**: Security and compliance are paramount in Salesforce development. ChatGPT can be tailored to understand an organization's specific compliance requirements and integrate those into the development process. Whether it's enforcing field-level security or ensuring that triggers handle bulk operations efficiently, ChatGPT can guide developers in aligning the code with best practices, thus ensuring robust, secure, and compliant solutions.

ChatGPT represents a valuable tool in the arsenal of Salesforce developers, particularly when working with triggers. By automating code generation, enhancing collaboration, assisting in testing, and ensuring adherence to best practices and compliance standards, it offers a pathway to faster and higher-quality development. It aligns well with the iterative and responsive nature of Salesforce development, facilitating a more efficient and effective approach to crafting robust and scalable triggers.

Designing triggers with ChatGPT's guidance

Let's start by defining a requirement that should be solved with a trigger. Whenever an account address is changed, the updated address needs to be sent to an ERP system. Assume that the Call Out has been written and the trigger simply needs to call the Call Out method. This would be an **after** trigger since what is being inserted/updated in the account record is not being changed. As a good design practice is to use a handler, that is what will be requested. For this example, there are not currently any account triggers:

Prompt	Please write a Salesforce after trigger to call a method whenever any of the address fields in the account record changes. Please write this using a trigger handler. Please write on insert and update.
Response	Certainly! Here's how you can structure a Salesforce after trigger for the account object to detect changes in address fields and then call a method when these fields are altered. This will include a trigger and a trigger handler class:

- `AccountTrigger.trigger`:

```
trigger AccountTrigger on Account (after insert,
after update) {
```

```
        if (Trigger.isAfter) {
            if (Trigger.isInsert || Trigger.isUpdate) {
                AccountTriggerHandler.
handleAddressChange(Trigger.new, Trigger.oldMap);
            }
        }
    }
```

- `AccountTriggerHandler.cls`:

```
public class AccountTriggerHandler {

    public static void
handleAddressChange(List<Account> newAccounts,
Map<Id, Account> oldMap) {
        List<Account> accountsWithChangedAddresses =
new List<Account>();

        for (Account acc : newAccounts) {
            if (oldMap == null ||
addressChanged(acc, oldMap.get(acc.Id))) {
                accountsWithChangedAddresses.
add(acc);
            }
        }

        if (!accountsWithChangedAddresses.isEmpty())
{
            callYourMethod(accountsWithChanged
Addresses);
        }
    }

    private static Boolean addressChanged(Account
newAccount, Account oldAccount) {
        return (newAccount.BillingStreet !=
oldAccount.BillingStreet ||
                newAccount.BillingCity !=
oldAccount.BillingCity ||
                newAccount.BillingState !=
oldAccount.BillingState ||
                newAccount.BillingPostalCode !=
oldAccount.BillingPostalCode ||
                newAccount.BillingCountry !=
oldAccount.BillingCountry ||
                newAccount.ShippingStreet !=
oldAccount.ShippingStreet ||
```

```
                    newAccount.ShippingCity !=
oldAccount.ShippingCity ||
                    newAccount.ShippingState !=
oldAccount.ShippingState ||
                    newAccount.ShippingPostalCode !=
oldAccount.ShippingPostalCode ||
                    newAccount.ShippingCountry !=
oldAccount.ShippingCountry);
        }

        private static void callYourMethod(List<Account>
accounts) {
            // Your logic here. This is the method that
gets called when any address field changes.
        }
}
```

This setup splits the logic from the trigger, keeping the trigger simple and pushing the bulk of the logic to the handler class. It checks for changes in both the billing and shipping addresses. If a change is detected, the `callYourMethod` function will be invoked. You can implement your logic inside this function as needed.

Note: Ensure you have test classes written for this to maintain Salesforce's code coverage requirements.

Upon analyzing ChatGPT's response, it's evident that the foundational structure of the trigger was constructed competently. The trigger accounted for record modifications and incorporated considerations for standard fields. It even included a reminder to create a corresponding test class. However, the quality of the code falls short of the standards required for actual project deployment. The most apparent areas for enhancement are the lack of inline comments and the absence of a comment header block. Now, let's request ChatGPT to integrate these comments along with the test class:

Prompt	Can you please rewrite the code using inline comments and a comment header block and include a test class? Please don't modify the code.

Response	Of course! I'll provide an improved version of the Salesforce trigger code and its handler, including a detailed comment header block, inline comments, and the outline for a test class:
	
	• `AccountTrigger.trigger`:
	
	`/**`
	`  * AccountTrigger`
	`  *`
	`  * This trigger detects address changes on the Account record.`

```
     * It is designed to operate after insert and after
update operations.
     *
     * Author: [Your Name]
     * Date: [Creation Date]
     */
trigger AccountTrigger on Account (after insert,
after update) {

     // After Insert/Update operations: Check for
address changes
     if (Trigger.isAfter && (Trigger.isInsert ||
Trigger.isUpdate)) {
          AccountTriggerHandler.
handleAddressChange(Trigger.new, Trigger.oldMap);
     }
}
```

- `AccountTriggerHandler.cls`:

```
/**
 * AccountTriggerHandler
 *
 * A handler class for the AccountTrigger. It houses
the logic to detect
 * address changes on Account records and calls
appropriate methods if changes are found.
 *
 * Author: [Your Name]
 * Date: [Creation Date]
 */
public class AccountTriggerHandler {

    /**
     * Processes the address change on Account
records.
     *
     * @param newAccounts - New versions of the
Account records
     * @param oldMap - Old versions of the Account
records for comparison
     */
    public static void
handleAddressChange(List<Account> newAccounts,
Map<Id, Account> oldMap) {
         List<Account> accountsWithChangedAddresses =
new List<Account>();
```

```
        // Iterate through the new Account records
        for (Account acc : newAccounts) {
            // If the record is a new insertion or
its address has changed, add to the list
            if (oldMap == null ||
addressChanged(acc, oldMap.get(acc.Id))) {
                accountsWithChangedAddresses.
add(acc);
            }
        }

        // If address changes are detected, call the
designated method
        if (!accountsWithChangedAddresses.isEmpty())
{
            processChangedAddresses(accountsWith
ChangedAddresses);
        }
    }

    /**
     * Compares address fields between old and new
Account records.
     *
     * @param newAccount - The updated Account
record
     * @param oldAccount - The original Account
record
     * @return - Returns true if there are changes,
otherwise false
     */
    private static Boolean addressChanged(Account
newAccount, Account oldAccount) {
        return (newAccount.BillingStreet !=
oldAccount.BillingStreet ||
                ... // [Other address fields
comparisons]
                newAccount.ShippingCountry !=
oldAccount.ShippingCountry);
    }

    /**
     * Method to process the Account records with
changed addresses.
     * Placeholder for the desired operations on
changed address records.
     *
```

```
        * @param accounts - List of Account records
   with address changes
        */
       private static void
   processChangedAddresses(List<Account> accounts) {
           // Logic to process the address changes
       }
   }
```

- `AccountTriggerTest.cls`:

```
   /**
    * AccountTriggerTest
    *
    * Test class for the AccountTrigger and its
   handler.
    * Ensures that address change detection and
   processing are functioning as expected.
    *
    * Author: [Your Name]
    * Date: [Creation Date]
    */
   @isTest
   private class AccountTriggerTest {

       @isTest static void testAddressChangeDetection()
   {
           // Setup: Create an Account record
           Account testAcc = new Account(Name='Test
   Account', BillingStreet='Old Street');
           insert testAcc;

           // Modify the BillingStreet field
           testAcc.BillingStreet = 'New Street';
           update testAcc;

           // [Assertions to validate that the address
   change processing logic was executed]
       }
   }
```

Remember to fill in the [Your Name] and [Creation Date] placeholders with appropriate values. The test class provided is a basic framework, and you'll want to expand upon it based on the specific processing you implement for address changes.

That's an improvement. With the inclusion of comments, test classes, and reminders to adjust certain comment sections, the code is now primed to be used as a template for your code in a new Salesforce project.

In this section, we explored how to leverage ChatGPT to generate Salesforce code templates, such as Apex triggers. But where do we go from here?

Optimizing and refactoring existing triggers

While ChatGPT is a powerful tool, it's not a replacement for formal code review processes, dedicated linting tools, or Salesforce's built-in development tools. Instead, it acts as a complementary resource to augment your development workflow and improve code quality. ChatGPT can be a valuable tool for Salesforce developers in several ways to ensure and improve the quality of their code. In this section, we will review the following two ways:

- **Best practices**:

  - **Guidance**: If you're unsure whether a particular approach aligns with Salesforce's best practices, ChatGPT can provide insights and suggestions

  - **Patterns and anti-patterns**: You can inquire about common design patterns and anti-patterns specific to Salesforce development

- **Performance optimization**:

  - **Query optimization**: ChatGPT can provide suggestions on how to optimize SOQL queries to avoid governor limits and improve performance

  - **Bulkification**: Ensure your code is bulkified to handle multiple records efficiently. If you're unsure how to bulkify certain logic, ChatGPT can assist

Let's consider the following example. Suppose we want to write a trigger that updates a `Custom_Status__c` custom field on the `Contact` object whenever the `Email` field changes.

Here's a non-bulkified trigger:

```
trigger ContactTrigger on Contact (before update) {
    // Loop through each Contact in the trigger context
    for (Contact newContact : Trigger.new) {
        // Retrieve the old version of the Contact for comparison
        Contact oldContact = Trigger.oldMap.get(newContact.Id);

        // Check if the Email field has changed
        if (oldContact.Email != newContact.Email) {
            // Query the Contact to get additional fields (non-
bulkified)
```

```
        Contact queriedContact = [SELECT Id, Custom_Status__c FROM
Contact WHERE Id = :newContact.Id LIMIT 1];

        // Update the custom field
        queriedContact.Custom_Status__c = 'Email Updated';

        // Update the Contact (non-bulkified)
        update queriedContact;
    }
  }
}
```

This code is not bulk-safe because of the following reasons:

1. It queries inside a loop, which means if the trigger processes multiple records simultaneously, you'll hit the SOQL query limit quickly

2.   It performs DML operations inside the loop, which again can quickly hit DML limits if you're processing multiple records

Now, let's ask ChatGPT to bulkify the code:

Prompt	Suppose we want to write a trigger that updates the Custom_Status__c custom field on the Contact object whenever the Email field changes.
	Could you please bulkify the following trigger?
| | ```
trigger ContactTrigger on Contact (before update) {
    // Loop through each Contact in the trigger context
    for (Contact newContact : Trigger.new) {
        // Retrieve the old version of the Contact for
comparison
        Contact oldContact = Trigger.oldMap.
get(newContact.Id);

        // Check if the Email field has changed
        if (oldContact.Email != newContact.Email) {
            // Query the Contact to get additional fields
(non-bulkified)
            Contact queriedContact = [SELECT Id, Custom_
Status__c FROM Contact WHERE Id = :newContact.Id LIMIT 1];

            // Update the custom field
            queriedContact.Custom_Status__c = 'Email
Updated';
``` |
```

<table>
<tr><td></td><td>

```
        // Update the Contact (non-bulkified)
        update queriedContact;
      }
    }
  }
```

</td></tr>
<tr><td>**Response**</td><td>

Certainly! Let's bulkify the given trigger.

To bulkify the trigger, we'll do the following:

1. Collect the IDs of the contacts where the email has changed.

2. Perform a single query outside the loop to get the required fields for all those contacts.

3. Modify the retrieved contacts.

4. Update the modified contacts in a single bulk update operation.

Here's the bulkified version of the trigger:

```
trigger ContactTrigger on Contact (before update) {

    // A list to store Contacts whose email has changed
and need to be updated
    List<Contact> contactsToUpdate = new List<Contact>();

    // Loop through each Contact in the trigger context to
identify ones with changed emails
    for (Contact newContact : Trigger.new) {
        Contact oldContact = Trigger.oldMap.
get(newContact.Id);

        // Check if the Email field has changed
        if (oldContact.Email != newContact.Email) {
            // Add the current contact Id to the list
(we'll query in bulk outside the loop)
            contactsToUpdate.add(new Contact(Id =
newContact.Id));
        }
    }
```

</td></tr>
</table>

```
        // Bulk query to get all contacts with changed emails
        if (!contactsToUpdate.isEmpty()) {
            List<Contact> queriedContacts = [SELECT
    Id, Custom_Status__c FROM Contact WHERE Id IN
    :contactsToUpdate];

            // Update the custom field for all queried
    contacts
            for (Contact c : queriedContacts) {
                c.Custom_Status__c = 'Email Updated';
            }

            // Bulk update operation
            update queriedContacts;
        }
    }
```

Now, the trigger performs only one SOQL query and one DML operation, irrespective of the number of contacts processed. This makes it scalable and efficient within Salesforce governor limits.

ChatGPT not only bulkified the trigger using a standard approach but also outlined the steps for educational purposes. This guidance can be invaluable for developers who are new to Salesforce coding. Additionally, based on prior feedback, pertinent comments were integrated into the code.

We can follow a similar approach with a more complex pattern – the Singleton pattern. The Singleton pattern is a design pattern that's used to restrict the instantiation of a class to a single object. This pattern ensures that a class has only one instance and provides a global point of access to that instance. It's particularly useful when exactly one object is needed to coordinate actions across the system. For example, you might have a database connection pool, a logging service, or a configuration manager that is best implemented using the Singleton pattern.

In object-oriented programming languages, the Singleton pattern is usually implemented by creating a class with a method that creates a new instance of the class if one doesn't exist. If an instance already exists, it returns a reference to that instance. To ensure that an instance is not created by another thread while the first thread is executing the code, the method is often marked as synchronized, or equivalent measures are taken.

The Singleton pattern can be advantageous for several reasons. First, it provides a single point of control, which can be advantageous for logging, driver objects, caching, thread pools, or database connections. It's also easy to subclass, and you can choose the instantiation process. However, it's worth noting that Singleton can be considered an anti-pattern in certain scenarios as it introduces global state into an application and can make code harder to test and reason about. So, while it can be incredibly useful, it should be implemented judiciously.

Let's consider the following code:

```
trigger AccountTrigger on Account (before insert) {
    // Loop through each Account in the trigger context
    for(Account acc : Trigger.new) {
        // Update the Description field
        acc.Description = 'Updated by trigger';
    }
}
```

This trigger doesn't use the Singleton pattern. It's straightforward and handles the task within the trigger itself. Now, let's consider the prompt and response:

Prompt	Could you rewrite the following Apex trigger so that it implements the Singleton pattern?
	
	`trigger AccountTrigger on Account (before insert) {`
	`    // Loop through each Account in the trigger context`
	`    for(Account acc : Trigger.new) {`
	`        // Update the Description field`
	`        acc.Description = 'Updated by trigger';`
	`    }`
	`}`

Response	Certainly! The Singleton pattern is often used in Salesforce to ensure that a particular class is instantiated only once during a single transaction. This can be useful for maintaining a single point of control, such as for caching or setting global variables.
	
	Here is how you could rewrite the trigger using the Singleton pattern:
	
	`// Singleton class to handle Account logic`
	`public class AccountTriggerHandler {`
	`    private static AccountTriggerHandler instance;`
	
	`    // Private constructor to prevent direct instantiation`
	`    private AccountTriggerHandler() {}`
	
	`    // Public method to get the instance`
	`    public static AccountTriggerHandler getInstance() {`
	`        if (instance == null) {`
	`            instance = new AccountTriggerHandler();`
	`        }`
	`        return instance;`
	`    }`

```
        // Method to update Account descriptions
        public void updateAccountDescriptions(List<Account>
accounts) {
            for(Account acc : accounts) {
                acc.Description = 'Updated by trigger with
Singleton pattern';
            }
        }
    }

    // Account trigger
    trigger AccountTrigger on Account (before insert) {
        AccountTriggerHandler handler = AccountTriggerHandler.
getInstance();
        handler.updateAccountDescriptions(Trigger.new);
    }
```

In this example, `AccountTriggerHandler` is the Singleton class that contains the logic for updating the account descriptions. The trigger itself (`AccountTrigger`) uses this Singleton class to perform the actual operation.

The previous prompt demonstrates several important concepts in Salesforce development, specifically regarding Apex triggers and design patterns:

- **Separation of concerns**: The trigger delegates the actual business logic to a separate handler class (`AccountTriggerHandler`). This makes the code easier to maintain and test.

- **Singleton pattern**: The handler class (`AccountTriggerHandler`) is implemented as a Singleton. This ensures that only one instance of the handler is used throughout a single transaction. This is particularly useful for scenarios where you need to maintain a state or cache data.

- **Encapsulation**: The Singleton class uses a private constructor and a public method to control its instantiation. This encapsulates the creation logic and ensures that outside classes can't directly instantiate it, adhering to good OOP principles.

- **Reusability**: By isolating the logic within a handler class, the code becomes more reusable. The same logic can potentially be invoked from other triggers or classes without duplication.

- **Readability**: The code becomes more readable when the business logic is separated from the trigger logic. This makes it easier for other developers to understand what the code is doing, which is particularly important in larger projects with multiple developers.

As we pivot from the task of refactoring Salesforce triggers to enhancing their performance and maintainability, it's critical to address the inevitable – debugging and error handling. Refactoring brings with it the promise of optimized functionality, but even the most elegantly refactored triggers are not immune to errors. The next section dives into the essential skills of troubleshooting: identifying the root causes of unexpected behavior, understanding the nuances of error messages, and resolving issues to ensure that our triggers not only perform efficiently but also reliably execute under all expected conditions. This part of our journey underscores the importance of resilience in trigger design, empowering us to navigate the complexities of debugging with confidence and precision.

Debugging and error handling in triggers

ChatGPT serves as an invaluable resource for debugging and error handling when working with Apex triggers in Salesforce. While Apex triggers are crucial for automating complex business processes, they can also introduce various challenges, such as unhandled exceptions, recursion issues, or governor limit breaches. ChatGPT can assist developers in identifying common pitfalls by analyzing code snippets, offering best practices for structuring triggers, and explaining how to employ defensive coding techniques such as try-catch blocks.

The utility of ChatGPT goes beyond merely pointing out syntax errors or runtime issues. It also provides contextual explanations and recommendations based on the Salesforce ecosystem's established guidelines. For instance, if a developer is struggling with bulkification issues or is unsure of how to manage trigger context variables effectively, ChatGPT can provide targeted advice along with code examples. This enables developers to not just fix their immediate issues but also to understand the "why" and the "how" behind effective Apex trigger development.

Furthermore, ChatGPT can guide developers through the creation of test classes that are essential for deployment and for ensuring the robustness of the trigger logic. By leveraging ChatGPT's capabilities, a developer can gain a more nuanced understanding of debugging and error-handling techniques, resulting in more resilient, efficient, and maintainable Apex triggers.

As a practical example, in Salesforce, you can't directly perform DML operations on `Trigger.new` or `Trigger.old` collections within the trigger. Attempting to do so will result in a runtime error. Here's an example to illustrate the issue:

```
trigger AccountTrigger on Account (before update) {
    delete Trigger.old;
}
```

In this example, the trigger tries to delete records from the `Trigger.old` list directly, which is not allowed and will result in a runtime error.

We can write a prompt to ask ChatGPT for help with the code. Here, it's assumed that I'm a junior developer and I'm not familiar with Apex:

Prompt	What is wrong with this trigger?

	`trigger AccountTrigger on Account (before update) { `
	`    delete Trigger.old;`
	`}`

Response	`delete Trigger.old;` is attempting to perform a delete DML operation directly on the `Trigger.old` list. This is not allowed in Salesforce Apex triggers. In Salesforce, `Trigger.old` and `Trigger.new` are read-only lists of sObjects, and you cannot perform DML operations on them directly.
	
	Specifically, consider the following:
	
	1. `Trigger.old` and `Trigger.new` are read-only in both "before" and "after" triggers.
	
	2. You cannot execute DML operations such as "insert," "update," or "delete" directly on these collections.
	
	If you want to delete records based on some condition during the "before update" trigger operation, you should put the IDs or instances of the records you want to delete in a separate list and then perform the delete operation on that list.
	
	Here's how you could rewrite the trigger to delete accounts:
	
	`trigger AccountTrigger on Account (before update) {`
	`    List<Account> accountsToDelete = new List<Account>();`
	
	`    for(Account acc : Trigger.old) {`
	`        // Apply your condition here, if any`
	`        accountsToDelete.add(new Account(Id = acc.Id));`
	`    }`
	
	`    if(!accountsToDelete.isEmpty()) {`
	`        delete accountsToDelete;`
	`    }`
	`}`
	
	In this rewritten version, a separate list called `accountsToDelete` is populated and then used for the delete operation. This way, you're not trying to perform DML directly on `Trigger.old`, and it will comply with Salesforce's rules.

ChatGPT has identified the issue, suggested a better approach, provided us with the code, and introduced Salesforce concepts. What a great help!

Summary

In this chapter, you expanded your proficiency in using Salesforce triggers and gained insights into their design, implementation, and optimization with the support of ChatGPT. This chapter provided a foundational understanding of triggers and demonstrated how ChatGPT's knowledge can guide the development of more efficient and robust triggers. You learned how to design and script triggers that align with business requirements and best practices, as well as to maintain existing triggers through refactoring for enhanced performance. Additionally, you acquired skills in troubleshooting, learning to quickly address and resolve trigger-related issues. Finally, this chapter instilled an appreciation for governance and best practices, ensuring you can maintain a well-organized and high-performing Salesforce environment.

In the next chapter, you will embark on a journey through the realm of **Lightning Web Components** (**LWC**), guided by the capabilities of ChatGPT. You will delve into the transformative power of LWC within Salesforce's UI, discovering how to leverage ChatGPT's insights to construct dynamic and sophisticated components. This chapter will cover the essentials of building, styling, and interacting with LWCs, as well as the critical techniques of troubleshooting and debugging. As you progress, you will develop a strong command over LWC development, from honing your proficiency in creating responsive UI elements to mastering the subtleties of user experience design. You'll also gain valuable skills in identifying and resolving LWC issues swiftly, ensuring your components meet the highest standards of quality and performance. Throughout the next chapter, the emphasis will be on continuous learning and adaptation, preparing you to stay abreast of the latest advancements in the fast-paced world of Salesforce UI development.

9

Using ChatGPT for Lightning Web Components

In recent years, Salesforce has embarked on a transformative journey, pivoting its UI development strategy with the introduction of **Lightning Web Components (LWC)**. At its core, LWC represents a convergence of modern web standards and Salesforce's proprietary ecosystem. Unlike its predecessor, the Aura framework, LWC introduces a component-based architecture that aligns with native web components' capabilities. This synchronicity with cutting-edge web standards means developers can seamlessly integrate the latest web development techniques, tools, and practices into their Salesforce projects. The pivot to LWC underscores Salesforce's dedication to fostering a UI that is not only interactive and user-friendly but also leverages the efficiency, modularity, and power that modern web technologies afford. As a result, developers equipped with LWC can sculpt richer, more engaging, and highly responsive user interfaces, setting new benchmarks in CRM user experiences.

ChatGPT – an unparalleled asset in LWC development

Navigating the intricacies of LWC can be a daunting task, especially for those accustomed to the Aura or Visualforce paradigms. This is where ChatGPT shines, offering unparalleled support in the LWC development journey. Its vast knowledge reservoir is adept at parsing complex LWC queries, providing instant feedback, elucidating best practices, and suggesting optimized code structures. More than just a reactive tool, ChatGPT is proactive; it anticipates common challenges developers might face and offers solutions before they morph into larger issues. For developers transitioning from other frameworks, ChatGPT demystifies the nuances of LWC, elucidating the architectural and syntactical differences and providing a roadmap for effective migration. The melding of ChatGPT's expansive knowledge base with LWC development ensures that both newcomers and seasoned developers can accelerate their coding, reduce debugging time, and maintain the highest standards of code quality.

Harmonizing ChatGPT's AI insights with Salesforce's LWC vision

The union of ChatGPT's AI capabilities with Salesforce's vision for LWC heralds a new era in CRM UI development. As LWC's capabilities expand and its features become more intricate, the significance of an intelligent, responsive tool such as ChatGPT is accentuated. ChatGPT serves as both a guide and a mentor, assisting developers in navigating the multifaceted world of LWC, while also instilling them with the confidence to innovate. It bridges gaps in knowledge, reduces the learning curve, and empowers developers to build with foresight. By comprehending the depth of LWC's transformative potential and harnessing ChatGPT's specialized insights, developers are not just keeping pace with Salesforce's UI evolution – they are at their vanguard, ready to sculpt next-generation, user-centric applications that redefine CRM engagements.

The structure of LWC

Understanding the structure of LWC is crucial for Salesforce developers. As LWC continues to grow in complexity and capability, having a detailed grasp of its architectural components becomes essential. The following points discuss the structure of an LWC:

- **Basics of the LWC architecture**: At its core, LWC is a module that encapsulates a reusable piece of web functionality. LWC's architectural design adheres to the Web Components standards, making it compatible with modern web development practices. This standardization allows developers to write code that's coherent with native HTML elements. The component-based structure promotes encapsulation, ensuring that each LWC is a self-sufficient entity, yet capable of interacting harmoniously with other components in a larger application ecosystem.

- **Files and folders in an LWC bundle**: When you create an LWC, it's not just a singular file; instead, it's a bundle of files, each serving a specific purpose. At a minimum, an LWC bundle consists of an HTML file, a JavaScript file, and a metadata configuration file. The HTML file defines the structure of your component – essentially, its user interface. The JavaScript file contains the logic for the component, dictating its behavior, handling user events, and managing data. The metadata file, which typically has a .xml extension, provides configuration details, such as the API version or where the component can be used within Salesforce. Beyond these core files, developers can also include other optional files, such as CSS for styling or additional resources for more complex functionalities. Let's look at an example:

- greetComponent.html (HTML file):

   This file defines the structure and user interface of the component:

```
<template>
    <lightning-card title="Greeting Card" icon-
name="custom:custom14">
        <div class="slds-m-around_medium">
            <lightning-input label="Enter your name"
value={userName} onchange={handleNameChange}></lightning-input>
            <p>{greetingMessage}</p>
```

```
        </div>
      </lightning-card>
    </template>
```

- `greetComponent.js` (JavaScript file):

This file provides the logic for the component:

```
import { LightningElement, track } from 'lwc';
export default class GreetComponent extends LightningElement {
    // Declare reactive variable
    @track userName = '';
    @track greetingMessage = '';

    // Handle user input and set the greeting message
    handleNameChange(event) {
        this.userName = event.target.value;
        if (this.userName) {
            this.greetingMessage = `Hello, ${this.userName}!`;
        } else {
            this.greetingMessage = '';
        }
    }
}
```

- `greetComponent.js-meta.xml` (metadata configuration file):

This XML file provides configuration details for the component:

```
<?xml version="1.0" encoding="UTF-8"?>
<LightningComponentBundle xmlns="http://soap.sforce.com/2006/04/
metadata" fqn="greetComponent">
    <apiVersion>52.0</apiVersion>
    <isExposed>true</isExposed>
    <targets>
        <!-- Specifies where the component can be used -->
        <target>lightning__AppPage</target>
        <target>lightning__RecordPage</target>
        <target>lightning__HomePage</target>
    </targets>
</LightningComponentBundle>
```

- For more complex LWCs, you may have additional files such as `greetComponent.css` for styling and static resources or other supporting files. This is a basic example, and real-world applications might involve more intricate functionalities, interactions with Salesforce data, integration with other components, and sophisticated user interfaces. However, this serves as a foundational blueprint to understand the structure of an LWC.

- **The role of decorators in LWC**: In the LWC JavaScript file, developers will often encounter **decorators**, which augment the properties and functions of the component. Some of the most common decorators are `@api`, `@track`, and `@wire`. The `@api` decorator makes properties and methods public, allowing them to be accessed by other components or set by the component's consumer. The `@track` decorator, on the other hand, was originally used to track changes to an object or array, ensuring the component re-renders when its internal state changes. However, with newer versions of LWC, primitive data types are reactive by default, reducing the need for this decorator. Lastly, the `@wire` decorator plays a crucial role in connecting a component to Salesforce data, often linking it to an Apex method or an object's records.

- **Templating and conditional rendering**: One of the powerful features of LWC is its dynamic templating system, which allows developers to render content conditionally. Using directives such as `if:true` or `if:false` in the HTML file, developers can control which parts of the component are displayed based on certain conditions. This facilitates the creation of responsive UIs that adapt to different data states or user inputs. Additionally, LWC supports iterative rendering using the `for:each` directive, enabling the component to display lists or arrays of items seamlessly. Combined with the data binding capabilities of the JavaScript class, this dynamic templating ensures that the component's UI remains in sync with its underlying data.

The structure and capabilities of LWC emphasize Salesforce's commitment to aligning with modern web standards, providing developers with a robust and intuitive framework to build efficient, responsive, and interconnected components.

This chapter intends to explain how ChatGPT can help with the overall LWC framework, making Salesforce developers more efficient and providing them with the learning material to grow in the role.

Building LWCs with ChatGPT's assistance

Salesforce LWCs can solve UI problems where the standard UI components cannot help. To see how ChatGPT can help our development team create an LWC, let's start with a user story.

User story – *create an LWC that allows us to view and edit all standard fields from the account and contact records in the same component*:

- **Goal**: As a sales rep, I want to create an account and contact record from a Salesforce component.

- **Acceptance criteria**:

  - **Given** a sales rep has navigated to the standard Salesforce Contact tab and selected a record

  - **When** the page is loaded

  - **Then** the sales rep can see account and contact records on the same page and edit the information

Let's get started:

Prompt	Could you please create an LWC for the following user story?
	
	User story – create an LWC that allows us to view and edit all standard fields from the account and contact records in the same component:
	
	    • **Goal**: As a sales rep, I want to create an account and contact record from a Salesforce component
	
	    • **Acceptance criteria**:
	
	        • **Given** a sales rep has navigated to the standard Salesforce Contact tab and selected a record
	
	        • **When** the page is loaded
	
	        • **Then** the sales rep can see account and contact records on the same page and edit the information

Response	Certainly! Creating an LWC to handle account and contact records together is a practical solution for Salesforce users such as sales reps who require quick access to these closely related entities. The LWC will display the fields and allow users to edit them. The design may vary depending on your specific needs, but the fundamental structure will include HTML, JavaScript, and an XML metadata file for the component.

Note: Salesforce has governor limits, so the example code is designed for educational purposes. You may need to adjust it so that it fits your real-world scenario:

- accountContactComponent.html:

```
<template>
    <lightning-card title="Account and Contact
Information">
        <!-- Account Section -->
        <div class="slds-m-around_medium">
            <h2>Account Information</h2>
            <template if:true={account}>
                <lightning-record-edit-form object-
api-name="Account" record-id={recordId}>
                    <lightning-messages></
lightning-messages>
                    <lightning-input-field field-
name="Name"></lightning-input-field>
                    <lightning-input-field field-
name="Phone"></lightning-input-field>
                    <lightning-input-field field-
name="Website"></lightning-input-field>
                    <lightning-button
variant="brand" type="submit" name="save"
label="Save Account"></lightning-button>
                </lightning-record-edit-form>
            </template>
        </div>
        <!-- Contact Section -->
        <div class="slds-m-around_medium">
            <h2>Contact Information</h2>
            <template if:true={contact}>
                <lightning-record-edit-form object-
api-name="Contact" record-id={recordId}>
                    <lightning-messages></
lightning-messages>
                    <lightning-input-field field-
name="LastName"></lightning-input-field>
                    <lightning-input-field field-
name="FirstName"></lightning-input-field>
                    <lightning-input-field field-
name="Email"></lightning-input-field>
                    <lightning-button
variant="brand" type="submit" name="save"
label="Save Contact"></lightning-button>
                </lightning-record-edit-form>
```

```
            </template>
          </div>
        </lightning-card>
    </template>
```

- accountContactComponent.js:

```
import { LightningElement, api } from 'lwc';
import { getRecord } from 'lightning/uiRecordApi';

export default class AccountContactComponent
extends LightningElement {
    @api recordId;
    account;
    contact;

    connectedCallback() {
        // Fetch the Account and Contact records
based on the given recordId (could be either
Account or Contact)
        // Implement this function to fetch record
details.
        // In real-world scenarios, you might
need server-side Apex classes to fetch complex or
related data.
    }
}
```

- accountContactComponent.js-meta.xml:

```
<?xml version="1.0" encoding="UTF-8"?>
<LightningComponentBundle xmlns="http://soap.
sforce.com/2006/04/metadata">
    <apiVersion>52.0</apiVersion>
    <isExposed>true</isExposed>
    <targets>
        <target>lightning__RecordPage</target>
    </targets>
</LightningComponentBundle>
```

This example is quite basic but meets the requirements of the user story to some extent. You may extend this by using server-side logic, validation, and more advanced features. For example, you can create an Apex class to serve as a controller to fetch account and contact data related to the given recordId. The example also assumes that the recordId value would be either an account or a contact, so you may need to implement logic to differentiate and handle this accordingly.

In this demonstration, ChatGPT provides a foundational, yet operational, example of an LWC designed to display and update both account and contact records on the same page. This example can serve as a starting point for more intricate use cases. As discussed in earlier chapters, the commenting could benefit from further refinement, and this example doesn't account for organizational specifics such as custom fields. However, it successfully interprets the core requirements of the user story and adeptly employs the LWC framework.

Let's see what happens if we provide some context and ask for header block comments.

Prompt	Could you generate the previous LWC again but this time include a custom field on the account named relationship__c and a header block comment per file?
Response	Certainly! Here is an example of the various files that would make up an LWC bundle for your specified user story. This bundle includes a custom field called relationship__c on the Account object and has a header block comment for each file to describe its purpose. • File 1 – myComponent.html: <!-- myComponent.html: This file is responsible for the component's structure and layout. -->

```
<template>
  <lightning-card title="Account and Contact
Information">
    <div class="slds-p-around_medium">
      <!-- Account Section -->
      <lightning-input label="Account
Name" value={account.Name}
onchange={handleAccountChange}></lightning-
input>
        <lightning-input label="Relationship"
value={account.relationship__c}
onchange={handleAccountChange}></lightning-
input>

        <!-- Contact Section -->
        <lightning-input label="Contact
Name" value={contact.Name}
onchange={handleContactChange}></lightning-
input>
        <lightning-input
label="Email" value={contact.Email}
onchange={handleContactChange}></lightning-
input>
      </div>
      <lightning-button label="Save"
onclick={handleSave}></lightning-button>
    </lightning-card>
</template>
```
|

- File 2 – myComponent.js:

// myComponent.js: This file contains the JavaScript logic controlling the component's behavior.

```
import { LightningElement, track } from
'lwc';

export default class MyComponent extends
LightningElement {
  @track account = {};
  @track contact = {};

  handleAccountChange(event) {
    const field = event.target.label;
    this.account[field] = event.target.value;
  }

  handleContactChange(event) {
    const field = event.target.label;
    this.contact[field] = event.target.value;
  }

  handleSave() {
    // Implement your saving logic here
  }
}
```

- File 3 – myComponent.js-meta.xml:

<!-- myComponent.js-meta.xml: This file specifies the API version and where the component can be used within Salesforce. -->

```
<?xml version="1.0" encoding="UTF-8"?>
<LightningComponentBundle xmlns="http://soap.
sforce.com/2006/04/metadata">
    <apiVersion>52.0</apiVersion>
    <isExposed>true</isExposed>
    <targets>
        <target>lightning__AppPage</target>
        <target>lightning__RecordPage</
target>
        <target>lightning__HomePage</target>
    </targets>
</LightningComponentBundle>
```

	Remember that this is a simplified example and that it doesn't include the backend logic to fetch or persist data in Salesforce. It focuses on the frontend LWC aspect and serves as a blueprint for more complex requirements.

This has improved the quality of the comments and included the desired field. With these two quick examples, we can expect various pros and cons when using ChatGPT in our development.

Here are the pros of using ChatGPT for LWC development:

- **Rapid prototyping**: ChatGPT can assist in quickly generating skeleton code or even more specific modules for LWC, enabling rapid prototyping and speeding up the initial development process.

- **Best practices**: The code that's generated can be aligned with Salesforce's best practices, helping novice developers understand the right way to structure their components, use decorators, and follow other best practices.

- **Design pattern guidance**: When facing complex architectural decisions, consulting with ChatGPT could yield design pattern recommendations, easing the decision-making process.

- **Code review assistance**: While it may not replace human expertise, ChatGPT can be an additional tool in code reviews, highlighting potential areas for optimization or improvement.

- **Problem-solving**: Got a particularly challenging bug or logic problem? ChatGPT can help brainstorm solutions or offer debugging tips, potentially reducing the time spent on troubleshooting.

Here are the cons of using ChatGPT for LWC development:

- **Limited context**: ChatGPT won't have the same understanding of your specific Salesforce org, custom objects, or business logic that a seasoned Salesforce developer would. This could result in solutions that are not fully aligned with your unique needs.

- **Complexity limitations**: While it can generate code snippets or components, ChatGPT's capacity to develop highly intricate, interconnected systems or understand deeply nested code logic is not equivalent to a human developer (yet).

- **Dependability**: While it can offer code suggestions, those suggestions are only as good as the latest training data and algorithms. They need to be validated and potentially modified by human experts to ensure they meet the project requirements.

- **No real-time collaboration**: ChatGPT cannot replace the benefits that come from real-time human collaboration such as pair programming, which is often essential for complex LWC development.

- **Limited to the frontend scope**: ChatGPT can help with frontend LWC code but will not be as proficient when it comes to integrating with complex backend logic or third-party services, limiting its utility to the full-stack nature of Salesforce projects.

In summary, while ChatGPT can offer substantial assistance in LWC development, it should be viewed as a complementary tool rather than a complete replacement for human expertise and context-aware decision-making.

Styling and interaction design

Utilizing ChatGPT as a resource for the styling and interaction design of LWCs can be an innovative way to streamline your development process. While generative AIs such as ChatGPT may not replace a seasoned UI/UX designer or frontend developer, they can provide invaluable guidance, tips, and even code snippets to help you navigate the nuanced landscape of LWC styling and interaction. Think of it as having a consultative resource at your fingertips, readily available to dispense targeted advice and solutions tailored to your specific design needs.

For starters, you can use ChatGPT to quickly mock up **Cascading Style Sheets** (**CSS**) classes that align with **Salesforce Lightning Design System** (**SLDS**), thereby ensuring a cohesive and native look and feel for your custom components. Simply ask the model to generate SLDS-compliant CSS code based on your requirements, and you'll have a solid foundation to build upon. The generated CSS can be a stepping stone, saving you the initial research and manual coding effort.

When it comes to interaction design, ChatGPT can assist in conceptualizing and implementing client-side logic. For example, you could ask for best practices on how to manage state in your LWC, how to implement drag-and-drop functionality, or even how to set up client-side routing. ChatGPT can provide sample JavaScript code that integrates smoothly with your LWC's HTML and CSS, as well as explain the rationale behind various design choices, making your development journey both educational and efficient.

Furthermore, ChatGPT can help identify possible UX pitfalls, offering suggestions on accessibility, responsive design, or browser compatibility issues you may need to consider. It can even aid in planning the user journey within your component, guiding you on where to place buttons, how to manage transitions, or when to use modals versus inline editing for a more intuitive user experience.

In essence, ChatGPT serves as a supplementary tool that can help accelerate your LWC styling and interaction design process, while also ensuring that you're adhering to best practices. It provides an extra layer of support that can refine your approach and potentially introduce you to new methods or techniques you hadn't considered.

Let's ask ChatGPT to generate the CSS mockup for the previous LWC:

Prompt	Can you please generate a SLDS-compliant CSS code for the previous LWC?
Response	Certainly! Here is an example of SLDS-compliant CSS code that could be used for an LWC that allows for viewing and editing of standard fields from account and contact records. This CSS assumes that the LWC has fields arranged in two separate sections or divs, one for account information and another for contact information:

```css
/* SLDS-Compliant CSS for the LWC */
/* General Component Styling */
.slds-card {
    padding: 1rem;
    border: 1px solid #e0e2e5;
    border-radius: 0.25rem;
    box-shadow: 0 2px 3px #ccc;
}
/* Account Section */
.account-section {
    background-color: #f3f2f2;
}
.account-section .slds-form-element {
    margin-bottom: 1rem;
}

.account-section .slds-form-element__label {
    font-weight: bold;
}

/* Contact Section */
.contact-section {
    background-color: #ffffff;
}

.contact-section .slds-form-element {
    margin-bottom: 1rem;
}
.contact-section .slds-form-element__label {
    font-weight: bold;
}

/* Custom Field */
.relationship-field {
    color: #0070d2;
}
```

```
/* Buttons */
.slds-button {
    margin-top: 1rem;
    margin-right: 0.5rem;
}
```

This example provides a separate look at the account and contact sections within the same page, with each section having its form elements styled differently for easy distinction. The .relationship-field class is used to style the custom relationship__c field on the account. Finally, the .slds-button class adjusts the button spacing.

Please adjust the classes as per the actual HTML structure of your LWC, though this should give you a good foundation to build upon.

With a single command, the CSS code was generated seamlessly, taking into account the context provided by previous instructions and aligning perfectly with Salesforce's SLDS guidelines.

In the next example, we'll ask for help implementing drag-and-drop client-side logic for an LWC where we want to categorize hot and cold leads in two columns. Let's have a look at the user story.

User story – *create an LWC that allows us to move leads with drag-and-drop capabilities from the hot column to the cold column to update the status field*:

- **Goal**: As a sales rep, I want to create an LWC to update the lead status from cold to hot, moving the lead from one column to another

- **Acceptance criteria**:

  - **Given** a sales rep has navigated to an LWC

  - **When** the page is loaded

  - **Then** the sales rep can see the list of hot leads and move them to the cold column

Let's take a look:

Prompt	Can you write the following LWC with drag-and-drop capabilities?
	- **User story** – create an LWC that allows us to move leads with drag-and-drop capabilities from the hot column to the cold column to update the status field:
	- **Goal**: As a sales rep, I want to create an LWC to update the lead status from cold to hot, moving the lead from one column to another

	• **Acceptance criteria**: ▪ **Given** a Sales Rep has navigated to an LWC ▪ **When** the page is loaded ▪ **Then** the sales rep can see the list of hot leads and move them to the cold column
Response	Certainly! To accomplish the drag-and-drop functionality described in your user story, we can make use of Salesforce's LWC, along with its native HTML drag-and-drop API. Here are the basic files you would require in your LWC bundle: • dragAndDropLead.html: This is your LWC HTML file:

```
<!-- dragAndDropLead.html -->
<template>
    <div class="container">
        <div id="hot-column" class="column"
ondrop={drop} ondragover={allowDrop}>
            <h2>Hot Leads</h2>
            <template for:each={hotLeads}
for:item="lead">
                <div key={lead.Id} class="lead-card"
draggable="true" ondragstart={drag}>
                    {lead.Name}
                </div>
            </template>
        </div>
        <div id="cold-column" class="column"
ondrop={drop} ondragover={allowDrop}>
            <h2>Cold Leads</h2>
            <template for:each={coldLeads}
for:item="lead">
                <div key={lead.Id} class="lead-card"
draggable="true" ondragstart={drag}>
                    {lead.Name}
                </div>
            </template>
        </div>
    </div>
</template>
```

- dragAndDropLead.js: This is your LWC JavaScript file:

```
// dragAndDropLead.js
import { LightningElement, wire, track } from 'lwc';
import getLeads from '@salesforce/apex/
LeadController.getLeads';
import updateLeadStatus from '@salesforce/apex/
LeadController.updateLeadStatus';

export default class DragAndDropLead extends
LightningElement {
    @track hotLeads = [];
    @track coldLeads = [];

    @wire(getLeads)
    wiredLeads({data}) {
        if(data) {
            this.hotLeads = data.filter(lead => lead.
Status === 'Hot');
            this.coldLeads = data.filter(lead =>
lead.Status === 'Cold');
        }
    }

    drag(event) {
        event.dataTransfer.setData("text", event.
target.dataset.id);
    }

    allowDrop(event) {
        event.preventDefault();
    }

    drop(event) {
        event.preventDefault();
        const leadId = event.dataTransfer.
getData("text");
        const targetColumn = event.currentTarget.id;

        // Update status based on the column
        let newStatus = targetColumn === 'hot-column'
? 'Hot' : 'Cold';
        updateLeadStatus({ leadId, newStatus });
    }
}
```

- dragAndDropLead.css**: This is your LWC CSS file:

```
/* dragAndDropLead.css */
.column {
    width: 300px;
    height: 500px;
    margin: 10px;
    padding: 10px;
    border: 1px solid black;
}

.lead-card {
    margin: 5px;
    padding: 10px;
    border: 1px solid gray;
}

.container {
    display: flex;
}
```

- LeadController.cls: This is your Apex class for interacting with Salesforce data:

```
// LeadController.cls
public with sharing class LeadController {
    @AuraEnabled(cacheable=true)
    public static List<Lead> getLeads() {
        return [SELECT Id, Name, Status FROM Lead
LIMIT 100];
    }

    @AuraEnabled
    public static void updateLeadStatus(Id leadId,
String newStatus) {
        Lead leadToUpdate = [SELECT Id, Status FROM
Lead WHERE Id = :leadId LIMIT 1];
        leadToUpdate.Status = newStatus;
        update leadToUpdate;
    }
}
```

This example assumes that you have a Lead object and a Status field on it. Feel free to adjust the fields and object names according to your Salesforce setup.

In the response, this time, ChatGPT has included client-side logic on the drag, drop, and allow drop functions to allow moving the lead from one column to another. Again, this is a basic example, where no context of the lead or any existing logic has been taken into consideration, but it could help with the prototype and manage expectations with the rest of the team members.

If we have a close look, it didn't come up with a fully customized solution. Instead, it's using the HTML API to align with standards on frontend development.

Overall, ChatGPT serves as an invaluable resource for delving into the capabilities of the LWC framework, assisting in both prototyping and blueprint creation for your Salesforce projects. By offering quick yet insightful solutions, it significantly accelerates the learning curve for new developers, helping them swiftly transition to senior-level Salesforce roles.

Troubleshooting and debugging LWC

ChatGPT offers a powerful avenue for troubleshooting and debugging issues within LWC in the Salesforce ecosystem. Armed with a vast repository of coding patterns, best practices, and common pitfalls, ChatGPT can aid in rapidly identifying code anomalies or logical inconsistencies that may be hindering your component's performance or functionality. For instance, if you're encountering a specific error message or odd behavior in your LWC, ChatGPT can help you interpret these indicators, guiding you toward the most likely sources of the problem.

Moreover, the platform's capability to understand complex technical prompts allows it to provide actionable insights. Instead of sifting through forums or documentation, you can receive immediate guidance on resolving issues such as data-binding problems, asynchronous call challenges, or difficulties with the component lifecycle. This real-time assistance not only speeds up the debugging process but also educates you on how to handle similar issues in the future.

Another significant advantage lies in ChatGPT's collaborative nature. Think of it as a mentor and a peer rolled into one – eager to share insights without overshadowing your problem-solving process. While it's not a replacement for in-depth debugging tools or thorough code reviews from experienced developers, it serves as an excellent first line of defense in troubleshooting. The nuanced blend of technical understanding and practical guidance makes ChatGPT an indispensable ally in debugging LWCs, allowing you to focus on crafting robust and efficient components.

So, whether you're new to Salesforce development or a seasoned pro looking to optimize your workflow, ChatGPT can be a transformative tool for elevating the quality of your LWC code, expediting the troubleshooting process, and imparting best practices for long-term success.

Here is an example of a simple LWC that contains an error. The component is supposed to display a list of account names fetched from Salesforce, but it contains an issue that will prevent it from working as expected:

- // myAccountList.js:

```
import { LightningElement, wire } from 'lwc';
import getAccounts from '@salesforce/apex/AccountController.
getAccounts';

export default class MyAccountList extends LightningElement {
    @wire(getAccounts) accounts;
    // Trying to log an Apex error message, but this is not the
right way to do it.
    errorMessage = this.accounts.error.body.message;
}
```

- <!-- myAccountList.html -->:

```
<template>
    <lightning-card title="Account List">
        <template if:true={accounts.data}>
            <template for:each={accounts.data}
for:item="account">
                <p key={account.Id}>{account.Name}</p>
            </template>
        </template>
        <template if:true={errorMessage}>
            <p>Error: {errorMessage}</p>
        </template>
    </lightning-card>
</template>
```

- // AccountController.cls:

```
public with sharing class AccountController {
    @AuraEnabled(cacheable=true)
    public static List<Account> getAccounts() {
        return [SELECT Id, Name FROM Account LIMIT 10];
    }
}
```

Error description: The issue in this LWC lies in the myAccountList.js file, specifically in the errorMessage = this.accounts.error.body.message; line.

Here, we are trying to capture an error message from the Apex method in case it fails. However, the way it's done is incorrect because the @wire property is not populated synchronously, and we can't access its error attribute in this manner.

This error might not be immediately evident during compilation, but it will throw a runtime error when you try to execute the component. The component will fail to render the error message properly if the Apex method encounters an issue. ChatGPT can identify the problem and help solve it with a quick prompt, as shown in the following example.

Prompt	Could you please explain why the following code is not working as expected?
	• // myAccountList.js:

	<pre>import { LightningElement, wire } from 'lwc';
	import getAccounts from '@salesforce/apex/
	AccountController.getAccounts';
	
	export default class MyAccountList extends
	LightningElement {
	    @wire(getAccounts) accounts;
	    // Trying to log an Apex error message, but this
	is not the right way to do it.
	    errorMessage = this.accounts.error.body.message;
	}</pre>

	• <!—myAccountList.html -->:

	<pre><template>
	    <lightning-card title="Account List">
	        <template if:true={accounts.data}>
	            <template for:each={accounts.data}
	for:item="account">
	                <p key={account.Id}>{account.Name}</
	p>
	            </template>
	        </template>
	        <template if:true={errorMessage}>
	            <p>Error: {errorMessage}</p>
	        </template>
	    </lightning-card>
	</template></pre>

	• // AccountController.cls:

	<pre>public with sharing class AccountController {
	    @AuraEnabled(cacheable=true)
	    public static List<Account> getAccounts() {
	        return [SELECT Id, Name FROM Account LIMIT
	10];
	    }
	}</pre>

Response	The code snippet you've posted has a mistake related to how it's attempting to handle errors when using the @wire decorator. The problem lies in this line:

```
errorMessage = this.accounts.error.body.message;
```

Here's a breakdown of what's wrong and why:

- The asynchronous nature of @wire: The @wire annotation fetches data or performs an operation asynchronously. So, trying to access properties of this. accounts.error right in the class body will not work as expected because this. accounts may not be initialized yet.

- **Error handling in @wire**: The proper way to handle errors in a @wire property is by using a JavaScript getter or utilizing a wired function to inspect the property.

Here's the corrected version of myAccountList.js:

```
import { LightningElement, wire, track } from 'lwc';
import getAccounts from '@salesforce/apex/
AccountController.getAccounts';

export default class MyAccountList extends
LightningElement {
    @wire(getAccounts)
    wiredAccounts({ error, data }) {
        if (data) {
            this.accounts = data;
            this.errorMessage = undefined;
        } else if (error) {
            this.errorMessage = error.body.message;
            this.accounts = undefined;
        }
    }

    @track accounts;
    @track errorMessage;
}
```

By using a wired function (wiredAccounts), you can correctly handle both the data and error scenarios. The @track decorator ensures that the component re-renders when these tracked properties change.

Your HTML can remain the same; it will now correctly display the accounts or an error message, depending on the state of the @wire call.

Spot on! ChatGPT has proven that it can provide feedback following the LWC terms and specific instructions and can create code that will help you fix the problem.

Summary

In this chapter, you've gained a comprehensive understanding of LWCs, laying a solid foundation for effective UI development in Salesforce. You've learned the skills needed to craft dynamic and responsive LWCs, with ChatGPT serving as a valuable guide throughout the development process. Additionally, this chapter has equipped you with the expertise to focus on design and user experience, emphasizing the importance of aligning LWCs with business branding for a seamless user interface. Finally, it has honed your ability to efficiently debug LWCs, enabling you to quickly resolve issues and maintain the high quality of your components.

In the next chapter, you'll delve into the crucial role of documentation in Salesforce projects and how ChatGPT can significantly enhance this process. You'll understand the importance of maintaining thorough documentation and learn how ChatGPT can be a valuable ally in creating and managing these vital resources. This chapter will guide you through drafting clear, concise, and comprehensive project documentation, including configuration notes and descriptions of custom code, with the assistance of ChatGPT. You'll also explore strategies to optimize existing documentation, using ChatGPT to identify and rectify any gaps, inconsistencies, or ambiguities. Furthermore, the next chapter will cover documenting best practices and guidelines, emphasizing compliance and quality, and leveraging ChatGPT for support. Finally, you'll learn about the role of ChatGPT in facilitating knowledge transfer and fostering collaborative documentation among team members, stakeholders, and successors, thereby ensuring a more inclusive and effective documentation process.

10

Using ChatGPT for Salesforce Project Documentation

In the ever-evolving landscape of Salesforce projects, one element remains constant and indispensable: documentation. It serves as the foundation upon which successful projects are built, providing a roadmap for developers, administrators, and stakeholders alike. Clear, well-maintained documentation is the bedrock of project success, ensuring continuity, collaboration, and effective knowledge transfer.

Writing documentation can be a daunting task for developers. That's where ChatGPT steps in as a valuable partner. ChatGPT can assist in formulating documentation that is not only clear and concise but also comprehensive. From capturing configuration notes to describing custom code intricacies, ChatGPT's generative AI capabilities can streamline the documentation process, enabling teams to focus on what matters most – delivering exceptional Salesforce solutions.

Furthermore, ChatGPT can unravel existing documentation. For reviewing pre-existing materials for gaps, inconsistencies, or ambiguities, or simply wanting to enhance the quality and relevance of the current documentation, ChatGPT can be a trusty companion.

This chapter will explore strategies to refine and elevate documentation to the highest standards of clarity, consistency, and compliance.

The importance of documentation and ChatGPT's role

High-quality documentation is the cornerstone of any successful Salesforce project, and its importance cannot be overstated. In the complex and dynamic world of Salesforce, where configurations, customizations, and integrations can be complex, well-maintained documentation serves as a vital resource. It acts as a comprehensive record of the project's design, development, and deployment, offering clarity and insight into every aspect of the solution. Without such documentation, the project becomes vulnerable to a host of challenges, including miscommunication, knowledge gaps, and inefficiencies, all of which can impede progress and hinder long-term success.

One of the primary roles of high-quality documentation is to ensure continuity. In Salesforce projects, team members may come and go, and stakeholders may change. Documentation serves as a timeless reference point that transcends personnel transitions, enabling new team members to quickly get up to speed and understand the project's intricacies. Additionally, it helps prevent knowledge silos by democratizing information, ensuring that critical insights and decision-making processes are not locked away in the minds of a few individuals. Moreover, documentation plays a pivotal role in risk mitigation as it provides a detailed history of changes and configurations, aiding in troubleshooting and debugging when issues arise. In essence, high-quality documentation is the key to maintaining project integrity, fostering collaboration, and ensuring the long-term success and sustainability of Salesforce projects.

ChatGPT can be a game-changer when it comes to crafting exceptional Salesforce project documentation. Leveraging its generative AI capabilities, ChatGPT can assist in several ways to ensure your documentation is of the highest quality. First and foremost, it can help in the initial drafting phase by generating clear, concise, and well-structured content. Whether you're documenting complex configurations, custom code, or business processes, ChatGPT can provide valuable input, helping you articulate your thoughts in a way that is easily understandable to both technical and non-technical stakeholders.

Another significant advantage of using ChatGPT for Salesforce project documentation is its ability to enhance consistency and adherence to best practices across the project team. It can provide guidance on following established standards and guidelines, ensuring that your documentation aligns with industry best practices and your organization's specific requirements. Moreover, ChatGPT can assist in maintaining up-to-date documentation by suggesting updates and revisions as your Salesforce project evolves. This ensures that your documentation remains a living resource that accurately reflects the current state of your project. Overall, ChatGPT can be a valuable partner in the documentation process, helping you save time, maintain consistency, and produce documentation that is both informative and accessible to all stakeholders involved in your Salesforce project.

Let's start by working with ChatGPT to create an outline of the **functional design document** (FDD) for a Salesforce project.

From the case study in the appendix, we have this project description:

"Acme Corp has recently deployed Salesforce Sales Cloud, and as part of that project created an integration with Oracle ERP. This integration keeps the Salesforce and Oracle ERP accounts in sync. Acme also has Snowflake in production; Snowflake is consuming data from multiple internal and external systems and has started to ramp up AI-based analytics. The Acme business leadership has asked the IT department to design a solution and estimate the time and cost of what it would take to have a "Snowflake" section on the Salesforce account pages that summarizes the information in Snowflake for the account and allows the sales rep to select a contact related to the account and automatically copy the Snowflake information to the email."

The first step is to ask ChatGPT to propose an outline for the FDD based on the project's description:

Prompt	Please write a detailed outline for the FDD for this project: Acme Corp has recently deployed Salesforce Sales Cloud, and as part of that project created an integration with Oracle ERP. This integration keeps the Salesforce and Oracle ERP accounts in sync. Acme also has Snowflake in production; Snowflake is consuming data from multiple internal and external systems and has started to ramp up AI-based analytics. The Acme business leadership has asked the IT department to design a solution and estimate the time and cost of what it would take to have a "Snowflake" section on the Salesforce account pages that summarizes the information in Snowflake for the account and allows the sales rep to select a contact related to the account and automatically copy the Snowflake information to the email.

Response	FDD outline – Snowflake integration with Salesforce Sales Cloud:
	
	1. I. **Introduction**:
	
	    A. A. **Project Overview**:
	
	        i. Brief description of Acme Corp's Salesforce Sales Cloud deployment and Oracle ERP integration.
	
	        ii. Introduction to Snowflake integration and its purpose.
	
	    B. **Project objectives**:
	
	        i. Clearly defined goals of the project.
	
	        ii. Expected outcomes and benefits.
	
	    C. **Scope of work**:
	
	        i. Explanation of the project's boundaries and limitations.
	
	        ii. Identification of the systems involved (Salesforce, Oracle ERP, Snowflake).
	
	    D. **The document's purpose**:
	
	        i. Clarify the purpose of this FDD.
	
	        ii. Set expectations for stakeholders and development teams.

2. **Snowflake integration with Salesforce Sales Cloud**:

   A. **Integration overview**:

      i. Detailed explanation of the proposed Snowflake integration with Salesforce Sales Cloud.

      ii. Diagram depicting the data flow between Salesforce, Snowflake, and other relevant systems.

   B. **Integration architecture**:

      i. Description of the technical architecture, including data connectors and APIs.

      ii. Data synchronization methodology.

   C. **Data mapping**:

      i. Mapping of Salesforce Account and Contact fields to Snowflake data.

      ii. Explanation of how data synchronization will occur.

   D. **User interface design**:

      i. Design concepts for the "Snowflake" section on Salesforce account pages.

      ii. User interface elements and layout.

      iii. Workflow diagrams for sales rep actions (copying Snowflake information to email).

   E. **Data security**:

      i. Measures to ensure data privacy and security.

      ii. Access controls and permissions.

   F. **Error handling**:

      i. Strategies for handling data synchronization errors.

      ii. Notification and logging mechanisms.

3. **Functional requirements**:

   A. **Salesforce account page enhancements**:

      i.   Detailed requirements for the "Snowflake" section on Salesforce account pages.

      ii. Functionalities for displaying and interacting with Snowflake data.

   B. **Contact-related Snowflake data**:

      i.   Requirements for linking contacts to Snowflake information.

      ii. Procedures for copying Snowflake data to emails.

   C. **Data synchronization**:

      i.   Detailed requirements for real-time or scheduled data synchronization.

      ii. Handling of data conflicts and updates.

   D. **Reporting and analytics**:

      i.   Considerations for reporting on Snowflake data within Salesforce.

      ii. Analytics capabilities for sales reps.

4. IV. **Non-functional requirements**:

   A. **Performance**:

      i.   Performance expectations for the Snowflake integration.

      ii. Response times and scalability considerations.

   B. **Scalability**:

      i.   Plans for accommodating future growth.

      ii. Scalability benchmarks.

   C. **Reliability**:

      i.   High availability and disaster recovery strategies.

      ii. Uptime requirements.

   D. **Compliance**:

      i.   Compliance with data regulations and standards.

      ii. Audit and monitoring requirements.

5. V. **Implementation plan**:

    A. **Development phases**:

        i.   Phased approach for implementing the Snowflake integration.
        ii. Milestones and timelines.

    B. **Resource allocation**:

        i.   Allocation of development teams, roles, and responsibilities.
        ii. Hardware and software requirements.

    C. **Testing strategy**:

        i.   Test scenarios and cases.
        ii. User acceptance testing plan.

    D. **Deployment plan**:

        i.   Procedures for deploying the integration to production.
        ii. Rollback plan in case of issues.

6. VI. **Cost estimation**:

    A. Detailed cost breakdown for development, testing, deployment, and ongoing maintenance.
    B. Budget allocation for hardware, software, and personnel.

7. VII. **Risk assessment**:

    A. Identification of potential risks and challenges.
    B. Mitigation strategies and contingency plans.

8. VIII. **Approval**:

    A. Sign-off section for key stakeholders to approve the FDD.

9. IX. **Appendix**:

    A. Supporting documents, diagrams, or reference materials.

This FDD serves as a comprehensive blueprint for the integration of Snowflake with Salesforce Sales Cloud, ensuring that all stakeholders have a clear understanding of the project's objectives, requirements, and implementation plan. It will guide the development teams and provide a basis for tracking progress and ensuring successful project delivery.

This FDD represents a robust foundation for the Snowflake integration with the Salesforce Sales Cloud project. It comprehensively outlines the project's objectives, scope, technical architecture, functional and non-functional requirements, implementation plan, cost estimation, risk assessment, and more. However, as a next step, it is essential to subject this FDD outline to a review process. This review will involve a thorough assessment of the organization's established standards and guidelines for documentation, ensuring that it aligns seamlessly with the company's specific requirements. Additionally, the review should draw upon the experience gained from past FDDs to ensure that all expected sections, details, and best practices are diligently incorporated into this document. This meticulous review process will not only enhance the FDD's completeness and coherence but also serve as a key quality assurance step on our path to a successful Snowflake integration with Salesforce Sales Cloud.

Drafting documentation with ChatGPT's assistance

Using ChatGPT to assist in writing documentation is an iterative process, and it is typically carried out section by section. Unlike simply instructing ChatGPT to generate an entire project FDD with a single prompt, breaking it down into sections allows for greater precision and control over the content. The process begins with creating a detailed outline, as we have done, which serves as the roadmap for the entire document. Once this outline is in place, the next step involves prompting ChatGPT for each specific section of the FDD individually. This approach ensures that each section receives the necessary attention and detail, allowing for a more thorough and coherent document overall. It also allows for collaboration and review between team members, who can collectively shape each section to meet the project's unique requirements. In essence, the iterative use of ChatGPT section by section enables a more controlled and customized approach to documentation, resulting in a high-quality final document that aligns precisely with the project's objectives and standards.

Using ChatGPT to write "*II. Snowflake integration with Salesforce Sales Cloud, A. Integration overview, 1. Detailed explanation of the proposed Snowflake integration with Salesforce Sales Cloud*" would start by navigating to the ChatGPT thread for the project:

Prompt	For the FDD for the "Snowflake integration with Salesforce Sales Cloud project," please write this section:
	II. Snowflake integration with Salesforce Sales Cloud, A. Integration overview, 1. Detailed explanation of the proposed Snowflake integration with Salesforce Sales Cloud.

Response	II. Snowflake integration with Salesforce Sales Cloud
	A. Integration overview
	The integration of Snowflake with Salesforce Sales Cloud represents a pivotal step in enhancing the capabilities of our Salesforce ecosystem. This section provides a detailed explanation of the proposed Snowflake integration, outlining the key components and functionalities that will be implemented to achieve seamless data synchronization and accessibility.
	The primary goal of this integration is to establish a bi-directional data flow between Snowflake and Salesforce Sales Cloud. This means that relevant data from Salesforce accounts and contacts will be synchronized with Snowflake, allowing for a centralized repository of information. Conversely, Snowflake will provide real-time or near-real-time data updates back to Salesforce, ensuring that our sales reps have access to the latest insights and analytics within the Salesforce interface.
	To achieve this, the integration architecture will leverage APIs and data connectors that facilitate secure and efficient data transmission between the two systems. Data mapping will be meticulously defined to ensure that the Salesforce Account and Contact fields are correctly associated with the corresponding Snowflake data.
	The integration will not only provide a comprehensive view of Snowflake data within Salesforce but also enable sales reps to interact with this data seamlessly. The "Snowflake" section on Salesforce account pages will serve as the central hub for accessing Snowflake insights, and sales reps will have the ability to select relevant contacts and effortlessly copy Snowflake information to emails. This integration will empower our sales team with data-driven decision-making capabilities, ultimately driving business growth and enhancing customer relationships.
	In summary, the Snowflake integration with Salesforce Sales Cloud is poised to revolutionize how we leverage data for sales and analytics. This section lays the foundation for a deeper dive into the technical aspects and functionalities of the integration, ensuring a clear understanding of its significance and potential impact on our organization.

It's important to note that ChatGPT, while a powerful tool for generating content, can occasionally employ hyperbole or overly enthusiastic language. In the context of documentation, especially for an FDD, it may be necessary to request that ChatGPT tone down any hyperbolic expressions to maintain a professional and objective tone. Documentation should prioritize clarity, accuracy, and a balanced presentation of information. Additionally, depending on the desired level of technical detail in the FDD, it is entirely feasible to instruct ChatGPT to include slightly more technical information when appropriate. While the FDD primarily focuses on functional aspects, a moderate amount of technical detail can provide valuable context and understanding without overwhelming the document. Striking the right balance between technical insights and accessible language is key to crafting an FDD that effectively communicates the project's goals, scope, and requirements.

Prompt	Great first pass at the FDD section content. However, please do three things: 1) rewrite it to employ less hyperbole and be more to the point with a project technical document flavor, 2) add some detail around the use of OAuth, REST, and JSON for the integrations, and 3) mention the use of a near real-time integration pattern.
Response	II. Snowflake integration with Salesforce Sales Cloud

A. Integration overview

The integration of Snowflake with Salesforce Sales Cloud is a key initiative that aims to enhance our data accessibility and decision-making capabilities. This section provides a concise overview of the proposed Snowflake integration, emphasizing technical details and integration methods.

Technical approach: This integration will rely on OAuth 2.0 for secure authentication and authorization between Salesforce and Snowflake. OAuth ensures that access to Snowflake data is controlled and granted only to authorized users and systems. RESTful APIs will be utilized for data exchange, providing a standardized and lightweight communication method. Data will be transmitted in JSON format, promoting compatibility and ease of data transformation.

Near real-time data synchronization: The integration pattern chosen for this project is near real-time data synchronization. Salesforce events, such as updates or inserts of account and contact records, will trigger data synchronization with Snowflake. Message queues or event streams will be employed to capture and transmit these events, ensuring minimal latency in data transfer.

Data flow: Data from Salesforce will be transformed to align with Snowflake's schema. Transformed data will then be loaded into Snowflake, either in real time or through scheduled batch processes. Bi-directional communication will also be established, allowing Snowflake to send updates or insights back to Salesforce when necessary.

This technical overview provides a clear picture of the integration's architecture and methodologies, highlighting the use of OAuth, REST, and JSON for secure and efficient data exchange. This near real-time integration pattern ensures that sales reps have access to the latest insights, contributing to informed decision-making within Salesforce Sales Cloud. |

It's important to acknowledge the dynamic nature of ChatGPT's outputs, which can vary significantly based on the prompts provided. Even slight differences in the wording or specificity of prompts can lead to distinct responses. In the context of an organization with established documentation standards, these standards should guide and shape the writing style of ChatGPT to ensure consistency and adherence to organizational norms. Detailed prompts that explicitly specify the desired tone, level of technical detail, and formatting are instrumental in obtaining responses that align precisely with the project's requirements.

Furthermore, it's worth emphasizing that the quality and relevance of ChatGPT's responses are directly proportional to the level of detail provided in the prompts. When prompts are comprehensive and well-defined, ChatGPT can generate responses that meet the organization's documentation needs effectively. Therefore, investing time and effort in crafting detailed prompts tailored to the organization's standards and the specific project's goals can yield more precise and valuable outputs, ensuring that ChatGPT becomes a valuable asset in generating high-quality documentation that aligns seamlessly with the project's objectives and the organization's standards.

Optimizing existing documentation

In the constantly evolving landscape of modern IT, the ability to adapt, enhance, and optimize existing documentation is critical. ChatGPT can be a lifesaver when used for document analysis and improvement. ChatGPT can be employed as a tool for evaluating, refining, and elevating an organization's existing documentation. Whether you're seeking to identify gaps, inconsistencies, or ambiguities within existing documentation, or simply looking to enhance its quality and relevance, this section guides you through the process of harnessing ChatGPT's capabilities to breathe new life into documentation efforts.

Starting with an example of lightweight documentation, let's explore how to use ChatGPT to improve the material:

- **Salesforce data integration to the data warehouse**: This documentation outlines the integration between Salesforce and our data warehouse to streamline data consolidation and analysis. The integration process involves transferring Salesforce data to our data warehouse for further reporting and analytics.

- **Data mapping**: Data mapping/transformation is performed to establish a link between Salesforce objects and corresponding data warehouse tables.

- **Authentication and authorization**: To ensure data security, an authentication mechanism is implemented. Salesforce and the data warehouse will exchange tokens for authentication purposes.

- **Data transfer**: Data transfer between Salesforce and the data warehouse is achieved using custom middleware written in C.

The first step is to ask ChatGPT to identify the gaps, inconsistencies, and ambiguities in the material.

Prompt	Please identify the gaps, inconsistencies, and ambiguities in the following material. Make suggestions on how to improve the material to enhance the project documentation containing the material.
	Salesforce data integration to the data warehouse: This documentation outlines the integration between Salesforce and our data warehouse to streamline data consolidation and analysis. The integration process involves transferring Salesforce data to our data warehouse for further reporting and analytics.
	Data mapping: Data mapping/transformation is performed to establish a link between Salesforce objects and corresponding data warehouse tables.
	Authentication and authorization: To ensure data security, an authentication mechanism is implemented. Salesforce and the data warehouse will exchange tokens for authentication purposes.
	Data transfer: Data transfer between Salesforce and the data warehouse is achieved using custom middleware written in C.
Response	Here are the gaps, inconsistencies, and ambiguities in the provided material, along with suggestions for improvement to enhance project documentation:
	1. **Lack of specifics in data mapping**:
	The section on data mapping is too brief and lacks specifics. It doesn't mention how data mapping is performed, what criteria are used, or provide any example mappings. To improve this section, provide more detailed information on how data mapping is conducted, including examples of Salesforce objects and their corresponding data warehouse tables.
	2. **Missing details in terms of authentication and authorization**:
	While the documentation mentions the implementation of an authentication mechanism, it doesn't specify which authentication method is used, how tokens are generated, or the security protocols in place. To enhance this section, provide clear details about the authentication process, token generation, and security measures employed to safeguard data during transfer.
	3. **Middleware technology not clearly defined**:
	The section on data transfer mentions the use of custom middleware written in C, but it lacks essential details about the middleware's configuration, functionality, and how it facilitates data transfer. To address this gap, provide a more comprehensive explanation of the middleware, its role, and how it ensures smooth data transfer between Salesforce and the data warehouse.

	4. **Structure and clarity**:
	The document lacks a clear structure with headings and subheadings, making it less organized and harder to follow. To enhance the documentation, introduce a clear structure with distinct sections for data mapping, authentication, data transfer, and any other relevant topics. Each section should provide detailed information and follow a consistent format.
	5. **Formatting and terminology**:
	Ensure consistent formatting and terminology throughout the document. For example, consider using a standardized format for headings and subheadings to improve readability. Additionally, use consistent terminology, such as "data transformation," rather than alternating between "data mapping" and "data transformation."
	By addressing these gaps, inconsistencies, and ambiguities and providing more detailed and structured information, the project documentation can become more informative and useful for the intended audience.

With input from the appropriate **subject matter experts** (**SMEs**), each of these issues can be addressed and the documentation updated.

Prompt	Please rewrite the following material to reflect the use of a) OAuth between Salesforce and the data warehouse, b) Salesforce custom web services and callouts, and c) Salesforce profiles, permission sets, and programmatic sharing for authorization in Salesforce.
	Authentication and authorization: To ensure data security, an authentication mechanism is implemented. Salesforce and the data warehouse will exchange tokens for authentication purposes.
Response	Authentication and authorization:
	To uphold robust data security, a comprehensive authentication mechanism is meticulously established for this integration. OAuth 2.0 is employed as the authentication protocol between Salesforce and the data warehouse, ensuring secure and standardized authorization. Salesforce custom web services and callouts facilitate this OAuth-based interaction, allowing for seamless communication between the two systems.
	For authorization within the Salesforce ecosystem, a combination of Salesforce profiles, permission sets, and programmatic sharing rules are meticulously configured. This approach ensures that access to sensitive data is strictly controlled and granted only to authorized users based on their roles and permissions. Profiles and permission sets define user privileges, while programmatic sharing rules enable fine-grained access control, aligning precisely with the project's security requirements and adhering to best practices.

Sitting with the SMEs and iterating through prompts and responses will result in a greatly improved document.

Prompt	What details can be provided to further improve this response?
Response	To further improve the response, you can provide additional details and context regarding the use of OAuth, Salesforce custom web services and callouts, and the specific configuration of Salesforce profiles, permission sets, and programmatic sharing for authorization. Here are some suggestions for enhancing the response:
	1. **OAuth configuration**:
	- Specify the OAuth version being used (for example, OAuth 2.0)
	- Describe the OAuth flow used in the integration (for example, authorization code flow, implicit flow)
	- Explain the process of obtaining OAuth tokens, including token request endpoints and grant types
	- Mention any token expiration and renewal mechanisms in place for long-running processes
	2. **Salesforce custom web services and callouts**:
	- Provide examples of custom web services or RESTful endpoints created in Salesforce for data communication
	- Explain how Salesforce callouts are used to interact with external systems, including the data warehouse
	- Mention any error handling and retry mechanisms for callouts to ensure data reliability
	3. **Salesforce authorization configuration**:
	- Detail the setup of Salesforce profiles, specifying the roles and responsibilities associated with each profile
	- Describe the creation of permission sets, including the specific permissions granted to users
	- Explain how programmatic sharing rules are defined, outlining the criteria and conditions for data access
	- Highlight any role hierarchy considerations in the authorization model

| | 4. **Security measures**:

- Discuss the security measures and best practices implemented to protect OAuth tokens during transmission and storage

- Explain how Salesforce's built-in security features, such as Shield Platform Encryption or Field-Level Security, are employed to enhance data security

- Mention any compliance standards or certifications (for example, HIPAA, GDPR, and so on) that the integration adheres to regarding data protection

5. **Logging and monitoring**:

- Describe the logging and monitoring mechanisms in place to track authentication and authorization activities

- Explain how administrators or security teams are notified of any suspicious or unauthorized access attempts

- Mention any audit trails or log retention policies for compliance and auditing purposes

Providing these additional details will offer a more comprehensive understanding of the authentication and authorization processes within the integration, promoting transparency and adherence to security best practices. |

Documenting best practices and guidelines

It's a bit meta to have ChatGPT write standards for what ChatGPT writes, but it's a glimpse into the true power and versatility of ChatGPT. In organizations fortunate enough to have well-established project documentation standards, there's a solid framework in place for creating clear, consistent, and high-quality documentation. These standards serve as a guiding light, ensuring that documentation aligns with the organization's objectives and maintains a cohesive style.

However, the reality is that project documentation standards may not always be up to date or, in some cases, may not exist at all. This is where ChatGPT can truly shine. ChatGPT can assist in creating, revising, or updating project documentation standards tailored to the organization's specific needs and goals. It can provide guidelines on formatting, tone, technical depth, and other critical aspects that ensure documentation remains effective and compliant with industry best practices. By leveraging ChatGPT to establish or refine documentation standards, organizations can benefit from improved clarity, consistency, and alignment with strategic objectives, ultimately leading to enhanced project communication and success.

Furthermore, ChatGPT's ability to adapt and learn from existing standards can help organizations stay ahead of the curve. It can analyze and incorporate industry-specific best practices and emerging trends into documentation standards, ensuring that the organization remains agile and well-prepared for evolving project needs. In essence, while established standards are invaluable, ChatGPT's capacity to create or update them offers a dynamic and forward-looking approach to project documentation that can significantly enhance the quality and effectiveness of documentation practices within the organization.

The first step is to get ChatGPT's take on the documents needed to define the standards for documents for information technology projects:

Prompt	What documents are needed to define the standards for documents for information technology projects?
Response	Defining standards for documents in information technology (IT) projects requires a comprehensive set of documentation standards. These standards help ensure consistency, clarity, and quality across various types of documents. Here are some essential documents that are needed to define these standards: 1. **Documentation standards guide**: A master document that outlines the overarching standards for all project-related documentation. It should cover formatting, style, language, and common elements such as headers, footers, and page numbering. 2. **Project documentation plan**: A document specifying which types of documents are required for a particular project, their purpose, and their intended audience. It should also include timelines for document creation and review. 3. **Template documents**: Templates for different types of project documents, such as project charters, requirements documents, design specifications, test plans, and more. These templates ensure that documents are consistent in structure and content. 4. **Style guide**: A guide that defines rules for writing, including grammar, punctuation, tone, and language preferences. It should specify how technical terms, acronyms, and abbreviations should be used. 5. **Version control policy**: Guidelines for version control, naming conventions, and document repository management. This ensures that the most current version of a document is always accessible and that changes are tracked. 6. **Review and approval process**: Procedures for document review, approval, and sign-off. This document outlines who is responsible for reviewing each document and the process for incorporating feedback and obtaining approvals. 7. **Documentation naming conventions**: A clear and consistent system for naming documents, files, and folders. This helps with the organization and retrieval of documents. 8. **Change control process**: Guidelines for making changes to documentation during the project life cycle. It should specify how changes are documented, reviewed, and approved.

	9. **Accessibility standards**: If applicable, standards for ensuring that documentation is accessible to individuals with disabilities, complying with relevant accessibility guidelines and laws. 10. **Security and confidentiality guidelines**: If sensitive information is included in project documents, guidelines for securing and restricting access to such documents. 11. **Archiving and retention policy**: Rules for archiving and retaining project documents after the project's completion, taking into account legal and compliance requirements. 12. **Training materials**: Materials to train project team members and stakeholders on how to adhere to the documentation standards and use the provided templates effectively. Collectively, these documents define a framework for creating, managing, and maintaining project documentation consistently and effectively. Adhering to these standards helps ensure that project documentation is of high quality, serves its intended purpose, and contributes to the success of IT projects.

Beyond the technology-agnostic documents are the technology-specific documents.

Prompt	This is a great list of technology-agnostic documents. Please provide a list of the documents specific to Salesforce projects (not a specific Salesforce project, but all Salesforce projects) where samples and templates would provide value – for example, Salesforce development standards, Salesforce delivery standards, and the like.
Response	Certainly! For Salesforce projects, there are specific documents and standards that organizations commonly use to ensure consistency and best practices. Here's a list of some essential Salesforce-specific documents and standards: 1. **Salesforce development standards**: This document outlines coding conventions, best practices, and guidelines for Salesforce development, covering aspects such as Apex, Lightning Components, Visualforce, and more. 2. **Salesforce data model design**: Details on how Salesforce objects, relationships, and fields are designed and structured within the organization's Salesforce instance. 3. **Salesforce security and access control standards**: Guidelines for setting up security roles, profiles, and permission sets to control data access and maintain data security. 4. **Salesforce deployment and release management standards**: Procedures for deploying changes from sandbox environments to production, including version control, continuous integration, and continuous deployment checklists. 5. **Salesforce integration standards**: Documentation on how integrations with other systems are designed and implemented, including API usage, data synchronization, and authentication methods.

6. **Salesforce testing standards**: Guidelines for planning, executing, and documenting testing activities, encompassing unit testing, integration testing, and **user acceptance testing (UAT)**.

7. **Salesforce Lightning Experience configuration standards**: Standards for configuring Lightning Experience, including page layouts, record types, and custom components.

8. **Salesforce reporting and dashboard standards**: Best practices for creating reports and dashboards to provide valuable insights to users.

9. **Salesforce documentation templates**: Templates for specific Salesforce project documents, such as requirements documents, data migration plans, and change management plans.

10. **Salesforce user training materials**: Materials for training Salesforce users, including end user guides, FAQs, and training videos.

11. **Salesforce data governance policies**: Policies and procedures for data management, data quality, and data retention within Salesforce.

12. **Salesforce compliance and regulatory standards**: Documentation to ensure that the organization's Salesforce implementation complies with industry-specific regulations and standards (for example, GDPR, HIPAA, and others).

13. **Salesforce Lightning Web Components development standards**: Guidelines for developing and deploying Lightning Web Components in Salesforce.

14. **Salesforce mobile app configuration standards**: Standards for configuring and optimizing Salesforce for mobile app usage, including mobile layouts and actions.

15. **Salesforce backup and recovery procedures**: Documentation on how Salesforce data and configurations are backed up and the procedures for data recovery in case of data loss.

These Salesforce-specific documents and standards are crucial for maintaining consistency, quality, and compliance within Salesforce projects. They serve as valuable resources for development teams, administrators, and stakeholders to ensure successful Salesforce implementations and ongoing management.

The next step for each of these documents is to work with ChatGPT to create the outline and populate the document section by section.

Knowledge transfer and collaborative documentation

The evolution of ChatGPT and similar AI models has been nothing short of remarkable. Today's ChatGPT, based on the GPT-4 architecture, represents the culmination of years of research and development in the field of natural language processing. It has demonstrated the ability to comprehend, generate, and manipulate text with a level of sophistication that was once considered science fiction. However, it's important to note that even though ChatGPT has reached an impressive level of capability, it continues to evolve rapidly, and its potential is far from fully realized.

One of the areas where ChatGPT's growth is most evident is in its capacity to consume, understand, and interact with documents. While there are token limits that define the amount of text it can handle in a single interaction, these limits are continually expanding. This means that ChatGPT can process increasingly longer documents and provide more contextually relevant responses. It's not just about the quantity of text but also the depth of comprehension. As the model continues to learn and adapt, its ability to accurately answer questions about documents and even generate coherent and informative updates to documents is constantly improving.

Looking ahead, the trajectory for ChatGPT and similar AI models is promising. As research and development in AI advances, we can expect these models to push the boundaries of what's possible. This means that the limits of what ChatGPT can do today, such as consuming, comprehending, and updating documents, will only become more expansive and sophisticated in the future.

One particularly exciting application of ChatGPT in the context of knowledge transfer is its role as a real-time collaborator and enhancer of documentation. During knowledge transfer sessions, whether it's onboarding new team members or sharing expertise within an organization, ChatGPT can play a pivotal role. It can assist in deep dives into complex topics, answer questions, and enhance documentation in real time. Imagine a scenario where SMEs and learners interact with ChatGPT during training sessions. ChatGPT can provide instant explanations, clarify doubts, and even suggest improvements to training materials or documentation, all while seamlessly integrating into the knowledge transfer process.

In essence, the synergy between human expertise and the evolving capabilities of ChatGPT holds great potential for accelerating knowledge transfer and enhancing documentation practices. As AI continues to advance, it will not replace human knowledge but rather augment and amplify it, creating a collaborative environment where humans and AI work together to achieve greater understanding and efficiency. This collaborative future is an exciting frontier, and the possibilities for leveraging ChatGPT in knowledge transfer and documentation enhancement are boundless.

Writing this section of this book was a challenging endeavor, primarily because the capabilities of ChatGPT, including its document consumption, explanation, and refactoring abilities, are continually evolving. As technology advances, it's challenging to provide a static snapshot that fully captures the dynamic nature of these capabilities. Therefore, we encourage you to log into ChatGPT and interact directly with the model to experience its current capabilities firsthand. By doing so, you can explore the latest features, ask specific questions, and witness the real-time document comprehension and enhancement capabilities in action. This hands-on approach allows for a more accurate and up-to-date understanding of what ChatGPT can achieve in the realm of document processing and knowledge transfer.

Summary

In this chapter, the role of documentation in Salesforce projects was explored, and the transformative power of ChatGPT as an invaluable ally was unveiled. This chapter laid the foundation for a documentation culture that thrives on clarity and continuity, demonstrating how ChatGPT can assist in crafting precise, comprehensive project documentation, from configuration notes to custom code descriptions, and fostering the fundamental skills needed for effective record-keeping. Additionally, this chapter delved into strategies for optimizing existing documentation, ensuring that it remains a reliable source of knowledge. By embracing documentation standards and best practices, you fortified your understanding of the principles that underpin effective documentation, all while harnessing ChatGPT to facilitate collaborative knowledge sharing and enhance team cohesion. As you move forward in your exploration of the power of generative AI in Salesforce projects, this newfound knowledge in documentation will serve as a strong pillar, supporting the success of your endeavors.

In the next chapter, you will delve into the pivotal role of user stories in Salesforce projects and discover how ChatGPT can be harnessed to enhance the process of crafting, refining, and managing these crucial narratives. The next chapter begins by exploring the core concept of user stories within Salesforce and highlights the advantages of integrating ChatGPT into the story creation and grooming process. Then, it delves into techniques for refining, decomposing, and prioritizing user stories, ensuring that the most critical business needs take precedence, all with the guidance of ChatGPT. You will also learn how to engage in a collaborative story-writing process, using ChatGPT to facilitate discussions, gather feedback, and create a shared understanding among stakeholders. Furthermore, the next chapter provides insights into staying agile and iterative in user story management, with ChatGPT offering guidance on best practices and iteration strategies. By the end of this chapter, you will have honed your abilities in crafting effective user stories, refining them efficiently, fostering collaboration, and embracing agile principles, all with ChatGPT as a valuable ally in the process.

11

Using ChatGPT for Salesforce User Stories

In the ever-evolving world of Salesforce, user stories stand as vital pillars that capture and convey the business needs and desires of stakeholders. These narratives lay the groundwork for targeted and efficient development, bridging the gap between abstract requirements and actionable tasks. This chapter describes an innovative approach to crafting, refining, and managing user stories by leveraging the power of ChatGPT.

Integrating ChatGPT into the delivery process can help confirm that user stories are not only accurately reflective of business needs but are also structured in a way that propels efficient development. This technology enhances clarity, fosters collaboration, and adds a nuanced understanding of priorities, facilitating an alignment between business goals and technological solutions.

Throughout this chapter, we will explore the concept of how ChatGPT can be employed as a potent tool in the creation and grooming process. From crafting impactful narratives to refining and prioritizing them, the insights provided here will act as a guide. Together, we'll uncover the collaborative dynamics and agile adaptability that ChatGPT brings to the table, thereby unlocking new horizons in the field of Salesforce customization and configuration. It's more than just an integration of technology; it's a pioneering approach that redefines the way we think about and execute Salesforce user-story definition.

Whether you're a developer, an architect, or a business analyst, the lessons and insights captured here will equip you with a robust understanding of how to innovate and excel in your projects using ChatGPT.

Understanding user stories and ChatGPT's contribution

User stories in the Salesforce ecosystem are essential tools for capturing end-user requirements. They are typically crafted through collaboration with stakeholders, particularly the product owner, and translate the users' needs into actionable functionalities. The conventional process involves detailed discussions, interpretations, and alignments with business goals to create concise and clear user stories. Working with the product owner, developers engage in a continuous dialogue to ensure user stories are both reflective of the users' needs and aligned with the product's vision.

Integrating ChatGPT into the process of creating and grooming user stories offers practical support for developers and product owners. By inputting specific contextual information about the project into ChatGPT, the AI can provide suggestions that aid in shaping clear and directly tied user stories. This assistance helps in refining the stories and aligning them with business objectives. The technology doesn't replace the human intellect in the process but rather augments it, offering insights that can lead to more precise user stories.

The collaboration between human decision-making and AI-driven insights such as ChatGPT brings a complementary dimension to the Salesforce user-story process. It assists in making iterative discussions with the product owner more focused, potentially reducing the time required to align stories with business values. While it doesn't revolutionize the process, the integration of ChatGPT can provide practical support in shaping and refining user stories, making the process more streamlined.

In sum, the combination of the traditional user-story creation process with the capabilities of ChatGPT provides an accelerated solution for developers and product owners. By adding AI-driven insights to human creativity, the creation and grooming of user stories can become a more efficient and targeted task. Let's take an in-depth look at how ChatGPT can contribute to the everyday practice of Salesforce professionals, offering a supportive role rather than a transformative one.

Refining and prioritizing user stories

The creation of an initial product backlog and user stories by the product owner is an essential starting point in Salesforce development. However, these user stories typically require further refinement, decomposition, and prioritization to become development-ready. This section focuses on practical techniques and approaches to shape these initial user stories into actionable tasks. It also explores how ChatGPT can provide guidance in this process, offering insights that may help in organizing and focusing on the most crucial business needs. This section aims to equip Salesforce professionals with the tools needed to transition from broad user stories to targeted development actions efficiently.

To illustrate this process, we are going to use a user story from the case study in *Appendix A* and utilize ChatGPT to assist in refining, expanding, and improving that user story.

Epic 2: Display of Snowflake Information on Salesforce Account Pages

User Story 2.1: **Create Snowflake Section within Salesforce Account Pages**

- **Goal**: As a Sales Rep, I want a dedicated "Snowflake" section within Salesforce Account pages so that I can quickly access the Snowflake data related to an account.

- **Acceptance Criteria**:

  - **Given** a Sales Rep is viewing a Salesforce Account page,

  - **When** the page is loaded,

  - **Then** a dedicated "Snowflake" section should be displayed.

Some common questions need to be addressed within the grooming process. Let's utilize ChatGPT to guide us along the next steps:

| Prompt | I've been given the below user story from the product owner. Please provide the initial next steps for how I should groom and refine this user story so that it meets INVEST criteria.

Epic 2: Display of Snowflake Information on Salesforce Account Pages

User Story 2.1: Create Snowflake Section within Salesforce Account Pages

- **Goal**: As a Sales Rep, I want a dedicated "Snowflake" section within Salesforce Account pages so that I can quickly access the Snowflake data related to an account.
- **Acceptance Criteria**:
  - **Given** a Sales Rep is viewing a Salesforce Account page,
  - **When** the page is loaded,
  - **Then** a dedicated "Snowflake" section should be displayed. |
|---|---|
| Response | Certainly! Grooming and refining the provided user story to meet the INVEST (Independent, Negotiable, Valuable, Estimable, Small, Testable) criteria will help to ensure that it's well understood, easily implemented, and delivers value to the users. Here's a step-by-step approach:

1. Independent: Check if the user story can be developed independently of other stories. If there are dependencies on other stories or epics, make sure to document them.

2. Negotiable: Discuss with the product owner and other stakeholders to confirm that there's flexibility in how the story can be implemented. This may require understanding different options for displaying the Snowflake data within Salesforce.

3. Valuable: Confirm that the goal and acceptance criteria align with the actual value the Sales Reps are expecting from having the "Snowflake" section. If necessary, gather more details on what specific Snowflake data they need and why.

4. Estimable: Break down the story into smaller tasks, such as designing the UI for the "Snowflake" section, integrating Snowflake data, and testing. Evaluate the complexity and provide an initial estimation of effort.

5. Small: If the story appears too large to be completed in a single iteration, consider splitting it into smaller, more manageable user stories. For example, one story could focus on the UI design, another on the integration, and so on. |

	6. Testable: The acceptance criteria provided are a good start but might need more details to create proper test cases. Collaborate with QA to define specific tests that verify the "Snowflake" section loads correctly and displays the expected data.
	7. Collaborate with the Development Team: Engage with your development team to discuss the technical feasibility and gather their insights on how best to implement the feature. Their input can greatly aid in refining the story.
	8. Validate with the Product Owner: Review the refined user story with the product owner to ensure that it aligns with their expectations and the needs of the Sales Reps.
	By following these steps, you'll help to align the user story with the INVEST criteria, providing a clear and actionable path for development. This refinement process emphasizes collaboration, understanding, and alignment with business values, promoting a more efficient and successful implementation.

ChatGPT provided the preceding response, which does not address our need as it chose to explain **INVEST** rather than provide the next steps for this specific user story.

Let's stay with ChatGPT and steer it in the right direction:

Prompt	Thank you. Rather than describing INVEST, please provide typical activities and questions I should be asking the product owner in order to groom and refine the user story.
Response	Certainly! Grooming and refining the user story requires a collaborative effort, and here's what you might typically do, including the key questions to ask the product owner
	Activities:
	Clarify the Objective: Ensure that you understand the purpose and benefit of the "Snowflake" section within the Salesforce Account pages.
	Questions: What specific Snowflake data needs to be displayed? Why is this data important for the Sales Reps?
	Understand the Acceptance Criteria: Go through the given criteria and ask for any additional requirements or specific conditions that need to be met.
	Questions: What should happen if the Snowflake data is not available? Are there any specific layout or formatting preferences?
	Assess Dependencies: Determine if there are any dependencies or constraints that need to be considered during development.
	Questions: Are there any dependencies on other epics or user stories? Are there any technical limitations or constraints to be aware of?

	Estimate the Effort: Discuss with the product owner the complexity of the story to ensure that it can be realistically implemented in a single iteration.

Questions: How critical is this feature, and when is it needed? Do you have any preference regarding the implementation approach?

Identify Potential Risks or Challenges: Identify any potential roadblocks or risks that might affect the implementation and discuss them with the product owner

Questions: Are there any known issues or challenges that might affect the implementation? What are your expectations for handling these challenges?

Review Mockups or Designs: If applicable, present mockups or preliminary designs for the "Snowflake" section and gather feedback.

Questions: Does this design meet your expectations? Are there any specific branding or visual guidelines that need to be followed?

Prioritize the Story: Understand the priority of this user story within the larger epic or project.

Questions: How does this user story rank in priority compared to others? Are there any deadlines associated with this feature?

Collaborate and Finalize: Iterate on the user story with the product owner making necessary adjustments until you both agree that it's ready for development.

Questions: Are there any additional details or clarifications needed? Are you satisfied with the final version of the user story?

By engaging in these activities and asking these targeted questions, you'll ensure that the user story is clear, actionable, and aligned with the product owner's expectations, setting the stage for successful development. |

The preceding response was better than the first response, but it still isn't quite what we need. Let's try again by clicking the **Regenerate** button in ChatGPT:

Prompt	<Regenerate> - Click the Regenerate button

Response	Certainly! Grooming and refining the user story requires specific activities and questions directed at the product owner to ensure clarity, alignment with business goals, and readiness for development. Here's what you might consider:
	Activities:
	1. Review the User Story: Start by analyzing the current state of the user story with your team to identify any areas that might need further clarification or details.
	2. Arrange a Meeting with the Product Owner: Schedule a focused session with the product owner to discuss the user story in depth.
	3. Collaborate with the Development Team: Encourage interaction between the product owner and the development team to ensure a shared understanding.
	4. Document Changes and Clarifications: Maintain a record of refinements made to the user story, capturing any additional requirements or changes.
	Questions to Ask the Product Owner:
	1. Understanding the Need:
	- "Can you explain more about why the Sales Reps need the "Snowflake" section?"
	- "What specific Snowflake data should be displayed, and how will it be used by the Sales Reps?"
	2. Clarifying the Acceptance Criteria:
	- "Are there specific visual or functional requirements for the "Snowflake" section?"
	- "How should the system behave if the Snowflake data is not available?"
	3. Exploring Dependencies:
	- "Are there any dependencies or prerequisites that we need to consider before implementing this?"
	- "Do we need to align this with any existing features or upcoming changes?"
	4. Evaluating Priority and Timeline:
	- "How critical is this feature, and when do you expect it to be delivered?"
	- "Are there any business events or dates that we should be aware of that might affect the priority or timeline?"

	5. Gathering Input on Potential Challenges:
	
	  - "Are there any known challenges or constraints that might impact the design or implementation of this feature?"
	
	6. Confirming the Details:
	
	  - "Have we captured all the necessary details, and does the refined user story align with your expectations?"
	
	By engaging in these activities and asking these targeted questions, you'll be working closely with the product owner to refine the user story. This collaborative approach ensures that the final user story is clear, aligned with business goals, and ready to be taken into development.

The preceding responses have provided some actionable ideas and next steps. As we progress on each of the next steps, we can also use ChatGPT to provide insights to enhance our discussions with the product owner and other stakeholders.

Let's return to ChatGPT:

Prompt	Please provide additional questions for discussion with the product owner within items 1, 2, and 3 of the "Questions to Ask the Product Owner" section.

Response	Certainly! Here are additional questions you can discuss with the product owner for those specific items:
	
	1. Understanding the Need:
	
	  - "What problem are we solving for the Sales Reps with the integration of the "Snowflake" section?"
	
	  - "How do you envision the Sales Reps interacting with the Snowflake data within the Salesforce Account pages?"
	
	  - "Are there specific user behaviors or scenarios we should consider while designing this section?"

2. Clarifying the Acceptance Criteria:

- "What formatting or presentation of the Snowflake data would be most useful for the Sales Reps?"

- "Is there a preferred location within the Salesforce Account pages for the "Snowflake" section?"

- "How should the system handle different levels of access or permissions for viewing the Snowflake data?"

- "Are there any compliance or security considerations we need to be aware of regarding the Snowflake data display?"

3. Exploring Dependencies:

- "Does this feature rely on any other upcoming or existing integrations with Snowflake or other systems?"

- "How does this user story align with or differ from similar features or sections within Salesforce?"

- "What external data sources, if any, will need to be considered for the Snowflake data integration?"

- "Are there any specific technical considerations or limitations we should be aware of while implementing this?"

These additional questions foster a more detailed and comprehensive discussion with the product owner By delving into these aspects, you ensure a well-rounded understanding of the requirements and expectations, leading to a user story that's more closely aligned with the actual needs and constraints of the project.

This response is better. It has given us a starting point for discussion with the product owner and other stakeholders within the user-story refinement process. As noted in the introduction to this chapter, ChatGPT isn't automating the entire user story refinement process, but it can significantly accelerate the process by providing us with valuable insights, inputs, and discussion topics along the way.

Now that we have a list of initial discussion points to utilize within the refinement process, let's move on to the collaborative story-writing process and see where ChatGPT can assist.

Collaborative story writing

Collaborative story writing is an essential aspect of the Salesforce project process. It involves key stakeholders such as the product owner developers, and business analysts working together to shape and refine user stories. In this section, we'll examine how tools such as ChatGPT can be utilized in the collaborative story-writing process, using the "Snowflake" section within Salesforce Account pages as a case in point.

ChatGPT can provide practical support in facilitating discussions, gathering feedback, and aligning understanding among the team members. The integration of such tools doesn't replace human insight but complements the team's efforts by offering structured guidance and suggestions. This approach aims to contribute to a more efficient and coherent story-writing process, helping to ensure that the final user stories accurately reflect the business needs and user requirements.

Here's how it manifests in three key areas:

1. **Facilitating discussions**: When team members discuss various facets of a user story, having a tool such as ChatGPT can help to moderate and guide the conversation. It can assist in keeping the discussions focused by suggesting relevant questions, highlighting potential dependencies, and offering insight into best practices. This guidance can be particularly useful when the team encounters complex issues or conflicts, as ChatGPT can provide unbiased perspectives based on industry standards and methodologies. It essentially acts as a knowledgeable facilitator, promoting productive dialogues.

2. **Gathering feedback**: ChatGPT's ability to analyze and synthesize information allows it to support the feedback collection process. Capturing the insights and concerns of different stakeholders can assist in organizing and summarizing the feedback, making it more accessible for review and consideration. For example, if multiple stakeholders have input on a particular feature or requirement, ChatGPT can compile these inputs into a coherent summary, highlighting areas of agreement or divergence.

3. **Aligning understanding**: Achieving alignment among team members is vital for a successful collaborative writing process. ChatGPT can contribute by providing clarification on terms, concepts, and methodologies that might be unfamiliar or ambiguous to some stakeholders. Through its capability to offer detailed explanations and examples, it helps bridge knowledge gaps and creates a shared understanding of the project's objectives and requirements. This alignment ensures that all team members are on the same page, enhancing collaboration and contributing to a more cohesive final product.

ChatGPT's role in the collaborative story-writing process isn't about automation but augmentation. It amplifies the team's capabilities, enriching discussions, streamlining feedback, and fostering alignment. This integration of technology doesn't replace human insight and collaboration but supports them, enhancing the team's ability to craft user stories that resonate with the project's goals and user needs.

Let's return to our case study example and walk through some examples of how to utilize ChatGPT within the collaborative story-writing process.

In our previous section, ChatGPT provided some initial questions to discuss with the product owner Let's assume that we have had a discussion with the product owner and now have additional input to provide to ChatGPT to further the writing process:

Prompt	The product owner and I had a discussion on the questions that were suggested related to understanding the need of this user story.
	Below are the answers to each question.
	Please update User Story 2.1 by incorporating these answers and then provide 3-5 additional questions to discuss with the product owner to further clarify or confirm the need.
	1. Understanding the Need:
	- Question: "What problem are we solving for the Sales Reps with the integration of the "Snowflake" section?"
	- Answer: "There is valuable data available within the Snowflake platform that is not visible to the Sales Reps today. Specifically, we would like our Sales Reps to be able to see Billing data, customer sentiment analysis, industry trends, and news related to the customer."
	- Question: "How do you envision the Sales Reps interacting with the Snowflake data within the Salesforce Account pages?"
	- Answer: "The data should be presented in an easily consumable format so that the Sales Rep can use it in real-time when interacting with a customer virtually or quickly prior to meeting a customer face to face. The Snowflake data is already summarized, so it just needs to be displayed nicely to the Sales Rep in Salesforce."
	- Question: "Are there specific user behaviors or scenarios we should consider while designing this section?"
	- Answer: "As mentioned already, we want the Sales Rep to be able to easily and quickly view the summarized Snowflake data within Salesforce. Also, we want the Sales Rep to be able to submit feedback on the accuracy of the data to Snowflake and Slack."
Response	Certainly! Based on the information you provided, the updated User Story 2.1 appears as follows:
	Epic 2: Display of Snowflake Information on Salesforce Account Pages
	User Story 2.1: Create Snowflake Section within Salesforce Account Pages
	Goal: As a Sales Rep, I want a dedicated "Snowflake" section within Salesforce Account pages so that I can quickly access Billing data, customer sentiment analysis, industry trends, and news related to the customer. The data must be presented in an easily consumable format to facilitate real-time interaction with customers, both virtually and in person.

Acceptance Criteria:

- Given a Sales Rep is viewing a Salesforce Account page,

- When the page is loaded,

- Then a dedicated "Snowflake" section should be displayed, summarizing Billing data, customer sentiment analysis, industry trends, and news related to the customer.

- And the Sales Rep must have the ability to submit feedback on the accuracy of the data to Snowflake and Slack.

Here are 3-5 additional questions to discuss with the product owner to further clarify or confirm the need:

1. Feedback Mechanism Specifics: "What specific mechanism should be used for the Sales Rep to submit feedback on the accuracy of the data? Do we need any specific integrations with Snowflake and Slack, and what will be the flow of this feedback process?"

2. Data Security and Compliance: "Are there any security or compliance considerations we need to take into account when displaying the Snowflake data within Salesforce? What measures should be put in place to ensure that sensitive information is handled appropriately?"

3. User Experience Considerations: "How should the Snowflake section be visually designed to align with the existing Salesforce Account page? Are there any specific guidelines or preferences for the layout, color scheme, or interactivity?"

4. Data Refresh and Synchronization: "How frequently should the Snowflake data be updated or synchronized within Salesforce? Are there specific requirements for real-time data retrieval or periodic refresh?"

5. User Roles and Permissions: "Are there different user roles that need specific access or restrictions to the Snowflake section? How do we ensure that the right people have the correct level of access?"

These questions delve into various aspects such as functionality, security, design, synchronization, and user access, which would help in refining the user story further to meet the exact needs of the stakeholders.

As demonstrated here, ChatGPT is enhancing our ability to facilitate the dialogue with the product owner In addition to accelerating discussion, ChatGPT has provided topics for discussion that we may have overlooked on our own.

Let's now consider hypothetical feedback that we have received from the product owner and business stakeholders after their review of the updated *User Story 2.1*. Rather than trying to address that feedback on our own, let's utilize ChatGPT to augment our process:

Prompt	The product owner and business stakeholders reviewed the updated User Story 2.1 and had the following feedback.
	Please synthesize this feedback, suggest a consolidated response to their feedback, and provide an updated User Story 2.1.
	Business Stakeholder 1: "Our Sales Reps don't have time to read a lengthy summary. Snowflake already provides this information in brief natural language sentences. We want that content visible in Salesforce rather than something new."
	Business Stakeholder 2: "Our Sales Reps need detailed, actionable data from Snowflake in order for this to be valuable. We should provide the Sales Rep with all the detailed data from Snowflake and they will know what to do with it."
	Product Owner: "We need to keep this simple. We have already invested the effort to generate summarized data within Snowflake."
Response	Certainly! It seems like there are differing opinions among the stakeholders, but the consensus is toward using the existing summarized data within Snowflake and keeping the integration simple. Here's the consolidated response to their feedback:
	1. Content Presentation: The integration will focus on displaying the existing summarized data from Snowflake, as generated. This ensures that the Sales Reps receive concise information without the need for additional processing.
	2. Balance between Summary and Detail: Recognizing the need for both brevity and detail, the integration will prioritize displaying the summarized information while exploring options for Sales Reps to access detailed data if required.
	3. Investment Consideration: Acknowledging the effort already invested in summarizing data within Snowflake, the integration will leverage this existing work and avoid duplication or reprocessing.
	Based on this consolidated understanding, here's the updated User Story 2.1:
	Epic 2: Display of Snowflake Information on Salesforce Account Pages
	User Story 2.1: Create Snowflake Section within Salesforce Account Pages
	Goal: As a Sales Rep, I want a dedicated "Snowflake" section within Salesforce Account pages, displaying the existing summarized Snowflake data so that I can quickly access information related to an account, including Billing data, customer sentiment analysis, industry trends, and news.

Acceptance Criteria:

- Given a Sales Rep is viewing a Salesforce Account page,

- When the page is loaded,

- Then a dedicated "Snowflake" section should be displayed, using the existing brief natural language sentences from Snowflake, summarizing Billing data, customer sentiment analysis, industry trends, and news related to the customer.

- And there should be an option for the Sales Rep to access more detailed data from Snowflake if needed.

- And the Sales Rep must have the ability to submit feedback on the accuracy of the data to Snowflake and Slack.

This updated version retains the original objectives while taking into consideration the stakeholders' emphasis on brevity, leveraging existing investments in summarizing data, and balancing the need for summarized and detailed information.

We can observe in the preceding example how ChatGPT has assisted us in synthesizing feedback from different stakeholders and providing a suggested conclusion or next step. While we could have done this manually, ChatGPT can accelerate the iterative user-story refinement process and potentially help us reach consensus faster among stakeholders.

Staying agile and iterative

The agile grooming process represents a crucial phase in the life cycle of a user story, transforming it from a raw draft penciled into an upcoming sprint into a refined narrative that meets the criteria of being "ready" for implementation. It begins with the introduction of a user story into the sprint backlog, often as a high-level idea or a customer need. During grooming sessions, the cross-functional team, including developers, testers, and product owners, collaboratively refines and details the story. This entails breaking it down into smaller, actionable tasks, defining acceptance criteria, and estimating its complexity. ChatGPT plays a pivotal role in this process by offering insights, facilitating discussions, and ensuring that the story aligns with best practices.

As grooming progresses, the user story evolves to a point where it becomes "ready" for development. This means that it is well-defined, with clear acceptance criteria that leave no room for ambiguity. It includes all the information necessary for the development team to understand what needs to be done and how to test it. ChatGPT's assistance in this phase aids in refining the story further, making sure that it meets the standard of being "ready" to enter the sprint. This transformation ensures that the user story seamlessly integrates into the agile development process, contributing to efficient and effective project execution.

From earlier in this chapter, there is *User Story 2.1: Create Snowflake Section within Salesforce Account Pages*:

Epic 2: Display of Snowflake Information on Salesforce Account Pages

User Story 2.1: Create Snowflake Section within Salesforce Account Pages

- **Goal**: As a Sales Rep, I want a dedicated "Snowflake" section within Salesforce Account pages, displaying the existing summarized Snowflake data so that I can quickly access information related to an account, including Billing data, customer sentiment analysis, industry trends, and news.

- **Acceptance Criteria**:

  - **Given** a Sales Rep is viewing a Salesforce Account page,

  - **When** the page is loaded,

  - **Then** a dedicated "Snowflake" section should be displayed, using the existing brief natural language sentences from Snowflake, summarizing Billing data, customer sentiment analysis, industry trends, and news related to the customer.

  - **And** there should be an option for the Sales Rep to access more detailed data from Snowflake if needed.

  - **And** the Sales Rep must have the ability to submit feedback on the accuracy of the data to Snowflake and Slack.

This story has been groomed and is close to ready for implementation. In a final review with the product owner, they (as product owners are prone to doing) want to add a feature to the user story. They ask to add a button to the "Snowflake" section that when clicked will email the information to the user:

Prompt	Please rewrite this user story to do everything it currently does as well as have a button that when clicked emails the Snowflake information to the current user:
	Epic 2: Display of Snowflake Information on Salesforce Account Pages
	User Story 2.1: Create Snowflake Section within Salesforce Account Pages
	• **Goal**: As a Sales Rep, I want a dedicated "Snowflake" section within Salesforce Account pages, displaying the existing summarized Snowflake data so that I can quickly access information related to an account, including Billing data, customer sentiment analysis, industry trends, and news.
	• **Acceptance Criteria**:
	• **Given** a Sales Rep is viewing a Salesforce Account page,
	• **When** the page is loaded,

	• **Then** a dedicated "Snowflake" section should be displayed, using the existing brief natural language sentences from Snowflake, summarizing Billing data, customer sentiment analysis, industry trends, and news related to the customer.
	• **And** there should be an option for the Sales Rep to access more detailed data from Snowflake if needed.
	• **And** the Sales Rep must have the ability to submit feedback on the accuracy of the data to Snowflake and Slack.
Response	Epic 2: Display of Snowflake Information on Salesforce Account Pages
	User Story 2.1: Create Snowflake Section within Salesforce Account Pages with Email Functionality
	• **Goal**: As a Sales Rep, I want a dedicated "Snowflake" section within Salesforce Account pages, displaying the existing summarized Snowflake data so that I can quickly access information related to an account, including Billing data, customer sentiment analysis, industry trends, and news. Additionally, I want the ability to email this Snowflake information to myself for easy reference.
	• **Acceptance Criteria**:
	• **Given** a Sales Rep is viewing a Salesforce Account page,
	• **When** the page is loaded,
	• **Then** a dedicated "Snowflake" section should be displayed, using the existing brief natural language sentences from Snowflake, summarizing Billing data, customer sentiment analysis, industry trends, and news related to the customer.
	• **And** there should be an option for the Sales Rep to access more detailed data from Snowflake if needed.
	• **And** the Sales Rep must have the ability to submit feedback on the accuracy of the data to Snowflake and Slack.
	• **And** a user-friendly "Email Snowflake Information" button should be available within the "Snowflake" section, allowing the Sales Rep to send the summarized Snowflake data to their email address for easy reference.
	This refined user story not only maintains the existing functionalities but also introduces a valuable addition: the "Email Snowflake Information" button. This button empowers Sales Reps to conveniently email the summarized Snowflake data to themselves, enhancing their ability to access and utilize this critical information.

ChatGPT's capacity to swiftly and efficiently aid in user-story grooming is nothing short of impressive. With its natural language processing (NLP) prowess and vast knowledge base, ChatGPT streamlines the process of refining and perfecting user stories. It assists teams in breaking down high-level ideas into actionable tasks, defining acceptance criteria, and estimating complexity. Its real-time feedback and guidance foster discussions among team members, ensuring a shared understanding of the story's nuances. ChatGPT's ability to identify gaps, ambiguities, and inconsistencies within user stories enhances their quality and sets the stage for more efficient and impactful implementation. In essence, ChatGPT acts as a valuable collaborator, accelerating the grooming process and contributing to the overall success of Salesforce projects.

Summary

This chapter has reviewed the pivotal role that user stories play in Salesforce projects and how ChatGPT can be a powerful ally in crafting, refining, and managing these narratives. It delved into the core concepts of user stories within Salesforce and explained the advantages of integrating ChatGPT into the story creation and grooming process, as well as shared techniques for refining, decomposing, and prioritizing user stories, ensuring that critical business needs take precedence, all with ChatGPT's guidance. ChatGPT has been used to facilitate discussions, gather feedback, and create a shared understanding among stakeholders.

The next chapter will delve into the critical phase of testing in Salesforce development and discover how ChatGPT can be seamlessly integrated into the testing process. It will explore fundamental principles of Salesforce testing and understand the unique value that ChatGPT brings to this domain and how to design clear, comprehensive, and relevant test cases for Salesforce development while benefiting from ChatGPT's guidance and expertise. The chapter will impart the ability to craft high-quality test scripts and proficiency in optimizing test scripts effectively, all with ChatGPT as a valuable resource.

12
Using ChatGPT for Salesforce Testing

Salesforce testing is pivotal in ensuring that the application developed meets the necessary quality, performance, and security standards. With Salesforce's role as a key platform for managing customer relationships and sensitive data, the reliability and robustness of the system are important. Thorough testing aids teams in discovering and addressing potential defects in the development process, leading to a more effective and trustworthy final product.

Salesforce testing serves as a safeguard for user experience and system stability. By systematically uncovering bugs and inconsistencies, testing ensures alignment with defined business requirements. Implementing rigorous testing procedures reduces the risk of defects post-deployment, saving both time and resources in the future. Testing confirms that the system functions appropriately under ideal conditions and holds up against unexpected or incorrect usage.

Positive and negative testing are two crucial aspects of Salesforce testing. Positive testing, often referred to as "happy path" testing, focuses on confirming expected behavior using valid input data. In contrast, negative testing employs invalid data to verify that the system can handle erroneous inputs gracefully. Both are vital for a complete validation of the system's functionality.

Structuring the testing scripts is a key aspect of this process. The test scripts should include the objective, preconditions, assumptions, step-by-step direction, and expected results. Clearly written test scripts help team members understand what, how, and why specific aspects are being tested. This serves as a roadmap for the entire project, ensuring a clear and cohesive approach to testing the system's various features.

Retesting plays a significant role in the QA process. After defects have been identified and corrected, retests are run under the exact same conditions to confirm that the specific issues have been resolved. This ensures that the changes haven't inadvertently created new problems, reinforcing the system's stability and functionality.

Preparing test users and data is essential to simulate real-world scenarios within Salesforce. Test users must be created with different profiles and permissions, mirroring the actual end users, while test data must accurately represent variations and edge cases. This preparation enhances the relevance of testing and offers insights into the system's real-world performance.

Salesforce testing transcends being mere procedural steps in the development cycle. It forms the backbone of practices that ensure system reliability, performance, and alignment with business goals. A well-thought-out approach to testing enhances the Salesforce instance's quality, engendering trust in its ability to support the organization's mission-critical operations.

ChatGPT's capabilities can be leveraged to significantly enhance the manual testing phase of Salesforce projects, offering a range of benefits.

In the area of test scripting, ChatGPT can provide guidance to test analysts, suggesting various scenarios to be tested, including both critical paths and edge cases. The platform's understanding of the complexities inherent in Salesforce and its various components enables it to assist in identifying specific areas that need to be carefully assessed. This ensures that testing is both comprehensive and meticulous, without overlooking essential functionalities.

Enhanced collaboration is another critical area where ChatGPT can make a difference. By acting as a bridge between business analysts, developers, and testing teams, and translating user stories in test scripts, it ensures a unified understanding of the intended functionality. This common understanding enables accurate testing and minimizes misunderstandings, reducing the need for unnecessary rework.

In addition to assisting with test scripting, ChatGPT can be leveraged as a valuable **just-in-time** (**JIT**) training tool for manual testers. By offering insights into best practices, common challenges, and effective testing strategies tailored to Salesforce's unique ecosystem, it supports continuous learning and skill enhancement. This educational aspect ensures that testers are well equipped to handle the nuanced demands of Salesforce testing.

ChatGPT offers a multifaceted support system for manual Salesforce testing efforts, from the scripting phase through hands-on testing. By providing nuanced guidance, fostering collaboration, aiding in the identification of issues, and supporting the continuous growth of testing skills, it adds significant value to the testing process. These advantages align with the overall objective of delivering high-quality Salesforce solutions, underscoring the potential of ChatGPT as a transformative tool in Salesforce development.

In this chapter, readers will delve into the critical role of testing in Salesforce development and learn how to integrate ChatGPT into this process for enhanced outcomes. The material covers the fundamentals of Salesforce testing, highlighting the nuances and the significant value ChatGPT brings to this domain. Participants will gain expertise in crafting effective test cases, using ChatGPT's guidance to develop clear, comprehensive, and relevant test scripts specifically tailored for Salesforce deployments. The chapter includes a practical case study, demonstrating the transition from user stories to Salesforce project test scripts. Key skills imparted include mastering the principles of effective Salesforce testing, developing high-quality test scripts, and achieving proficiency in optimizing these scripts within the Salesforce environment, all augmented by ChatGPT's insights. This comprehensive approach ensures that learners are equipped to produce Salesforce deployments that are not only functionally robust but also reliable and user-friendly.

Crafting effective test cases with ChatGPT

Using ChatGPT to write manual test scripts for Salesforce user stories is a seamless process that can drastically enhance the efficiency and precision of your testing phase. First, it involves understanding the user stories, goals, and acceptance criteria in detail. By inputting these elements into ChatGPT, the AI can then assist in defining a range of scenarios that cover both primary pathways and edge cases tailored to Salesforce's environment. Through an iterative collaboration between the development team and ChatGPT, testers can quickly translate complex user stories into detailed manual test scripts. This not only streamlines the process but also ensures that the scripts are aligned with business goals and compliance requirements specific to Salesforce, thereby integrating ChatGPT into the team's testing strategy as a valuable and insightful tool.

Unlike some facets of Salesforce development, there are no industry standards for test scripts. Fortunately, ChatGPT can help.

Prompt	What is the best way to document a Salesforce project test script?

Response	The best way to document the steps in a Salesforce project test script is to employ a structured, comprehensive approach that aligns with both the technical and business requirements of the test scenario.
	1. Title and Objective: Begin by defining a clear title and objective that aligns with the specific user story or requirement being tested. This sets the context and allows anyone reviewing the script to quickly understand its purpose.
	2. Preconditions and Assumptions: Outline any prerequisites or environmental conditions that must be in place before testing, such as specific user roles, data setup, or configuration settings within Salesforce.
	3. Test Steps and Expected Results: Detail each step of the process in a sequential manner, incorporating all actions that need to be performed. Each step should include the action, input data, and expected result, providing a clear pathway for execution. Utilizing tables or numbered lists can enhance readability.
	4. Negative Scenarios and Edge Scripts: Include provisions for negative testing and edge scripts, specifying how the system should respond to invalid inputs or unexpected user behavior.
	5. Attachments and References: If there are supporting documents, such as screenshots, flow diagrams, or references to specific Salesforce objects, include them as attachments or links.
	6. Version Control and Author Information: Implement version control to track changes and adaptations to the test script. Include details of the author, review status, and any relevant comments.
	7. Integration with Test Management Tools: If the team uses test management tools, the script should be structured to facilitate integration, enabling better tracking, reporting, and collaboration.
	By focusing on clarity, thoroughness, and alignment with the specific Salesforce functionality being tested, the test script becomes a robust tool, not just for QA professionals but for developers, analysts, and stakeholders, enhancing the quality and effectiveness of the testing phase.

Based on the response from ChatGPT, for each user story, ChatGPT can help write the following:

- Positive, negative, and edge test scripts

- For each test script:

  - Title and objectives

  - Preconditions and assumptions

  - Test steps and expected results

  - Test data

This information can be used to create a standard test script prompt that can be combined with a user story and any needed refining detail:

Please write at least one positive test script, at least one negative test script, and at least one edge test script for the user story below. Include a title, an objective, preconditions, assumptions, and detailed step-by-step directions. Include the expected results for each step.

Case study

Let's combine the test script prompt with a basic user story from *Appendix A*.

Prompt	Please write at least one positive test script, at least one negative test script, and at least one edge test script for the user story below. Include a title, an objective, preconditions, assumptions, and detailed step-by-step directions. Include the expected results for each step.
	Epic 2. Display of Snowflake Information on Salesforce Account Pages
	User Story 1: Create Snowflake Section within Salesforce Account Pages
	• **Goal**: As a Sales Rep, I want a dedicated "Snowflake" section within Salesforce Account pages so that I can quickly access the Snowflake data related to an account.
	• **Acceptance Criteria**:
	• **Given** a Sales Rep is viewing a Salesforce Account page,
	• **When** the page is loaded,
	• **Then** a dedicated "Snowflake" section should be displayed.

Response	Certainly! Below are the test scripts for positive, negative, and edge cases based on the provided user story.
	Positive Test Script
	Title: Verify the Display of the Snowflake Section within Salesforce Account Pages
	Objective: Ensure that the dedicated "Snowflake" section appears correctly when a Sales Rep views a Salesforce Account page.
	Preconditions: Sales Rep is logged in to Salesforce, and the Snowflake section is already implemented.
	Assumptions: The Sales Rep has appropriate permissions to view the Account page.
	Steps:
	1. **Navigate to Account Page**: Open an existing Salesforce Account page.
	• Expected Result: The Account page loads successfully.
	2. **Verify Snowflake Section**: Check if the dedicated **Snowflake** section is displayed on the page.
	• Expected Result: The **Snowflake** section is displayed within the Account page.
	Negative Test Script
	Title: Verify the Absence of the Snowflake Section for Non-Sales Rep User
	Objective: Ensure that users without Sales Rep permissions do not see the Snowflake section within Salesforce Account pages.
	Preconditions: A non-Sales Rep user is logged in to Salesforce, and the Snowflake section is implemented.
	Assumptions: The non-Sales Rep user has limited permissions that exclude access to the Snowflake section.
	Steps:
	1. **Navigate to Account Page**: Open an existing Salesforce Account page as a non-Sales Rep user.
	• Expected Result: The Account page loads successfully.

2. **Verify Absence of Snowflake Section**: Check if the dedicated **Snowflake** section is absent on the page.

- Expected Result: The **Snowflake** section is not displayed within the Account page for the non-Sales Rep user.

Edge Test Script

Title: Verify the Display of the Snowflake Section with Slow Internet Connection

Objective: Validate the display of the Snowflake section within Salesforce Account pages when the user has a slow internet connection.

Preconditions: Sales Rep is logged in to Salesforce, the Snowflake section is implemented, and the internet connection is intentionally slowed down (can be simulated using browser tools).

Assumptions: Sales Rep has proper permissions, and the system can handle varying internet speeds.

Steps:

1. **Navigate to Account Page**: Open an existing Salesforce Account page with a slowed internet connection.

- Expected Result: The Account page takes a longer time to load but eventually loads successfully.

2. **Verify Snowflake Section**: Check if the dedicated "Snowflake" section is displayed on the page, even with a slow connection.

- Expected Result: The "Snowflake" section is displayed within the Account page, although it might take more time to load compared to a regular connection.

These test scripts cover different aspects of the same functionality and together ensure that the feature behaves correctly under various scenarios, thereby contributing to robust and reliable software.

Writing detailed test scripts requires a thoughtful examination of the functionality to be implemented. Test scripts ensure alignment, spanning business analysts, developers, testers, and stakeholders. This alignment is vital, as it allows potential issues to be identified during the development phase, thus minimizing the risk of costly rework later in the project. Carefully designed test cases covering positive, negative, and edge scenarios provide a comprehensive perspective, ensuring that the implementation aligns with the intended functionality.

The step-by-step instructions in the test scripts create a precise roadmap for validation. By systematically verifying each aspect of the requirement, the tester can ensure that functionality works as expected, contributing to the overall quality of the feature. This detailed approach helps in confirming that the test is executed properly.

The test cases can also serve as a reusable asset, supporting future development, regression testing, and the conversation to automating testing.

This first example provides good test scripts for what are basic user stories and from a **user interface** (**UI**) POV simply display some additional fields.

A more complex scenario from *Appendix A* allows the user to send an email from the Account page. This feature spans five user stories. The first user story adds an **Email** button to the customer Lightning Web Component that displays Snowflake data.

Prompt	Please write at least one positive test script, at least one negative test script, and at least one edge test script for the user story below. Include a title, an objective, preconditions, assumptions, and detailed step-by-step directions. Include the expected results for each step.
	Epic 3. **Email Functionality with Snowflake Data**
	User Story 1: **Add an "Email" Button to the Snowflake Lightning Web Component**
	• **Goal**: As a Sales Rep, I want an "Email" button within the Snowflake section of the Account page so that I can initiate the process of sending Snowflake information via email.
	• **Acceptance Criteria**:
	▪ **Given** a Sales Rep is viewing the Snowflake section on the Account page,
	▪ **When** the Sales Rep looks at the Snowflake section,
	▪ **Then** an "Email" button should be visible and clickable.

Response	**Positive Test Script**
	Title: Verify "Email" Button in Snowflake Section for Sales Rep
	Objective: To validate the presence and functionality of the **Email** button within the Snowflake section of the Account page for Sales Reps.
	Preconditions: The Sales Rep is logged in to Salesforce and is on an Account page with a Snowflake section.
	Assumptions: The Snowflake section is functioning properly, and the email feature has been properly configured.
	Steps:
	1. **Verify Account Page**: Check that the Account page has loaded successfully.
	• Expected Result: Account page displays with all relevant sections, including the Snowflake section.
	2. **Inspect Snowflake Section**: Scroll to the Snowflake section and look for the "Email" button.
	• Expected Result: The **Email** button is visible within the Snowflake section.
	3. **Click on Email Button**: Click the **Email** button to test its clickability.
	• Expected Result: The button is clickable and initiates the email process as configured.
	Negative Test Script
	Title: Verify Absence of **Email** Button for Non-Sales Rep
	Objective: To ensure that the **Email** button within the Snowflake section is not visible or accessible to non-Sales Rep users.
	Preconditions: A non-Sales Rep user is logged in to Salesforce and is on an Account page with a Snowflake section.
	Assumptions: The feature is restricted to Sales Rep users only.
	Steps:
	1. **Verify Account Page**: Check that the Account page has loaded successfully.
	• Expected Result: Account page displays without Sales Rep-specific features.

2. **Inspect Snowflake Section**: Scroll to the Snowflake section and look for the "Email" button.

- Expected Result: The "Email" button is not visible within the Snowflake section for non-Sales Rep users.

Edge Test Script

Title: Verify **Email** Button with Incomplete Snowflake Data

Objective: To validate the behavior of the **Email** button within the Snowflake section when the Snowflake data is incomplete or not fully loaded.

Preconditions: The Sales Rep is logged in to Salesforce and is on an Account page where the Snowflake data is not fully loaded.

Assumptions: The Snowflake section and **Email** button are properly configured, but the Snowflake data source is experiencing delays or has incomplete data.

Steps:

1. **Verify Account Page with Incomplete Data**: Check that the Account page has loaded with incomplete Snowflake data.

- Expected Result: Account page displays with the Snowflake section showing incomplete or loading data.

2. **Inspect Snowflake Section**: Scroll to the Snowflake section and look for the "Email" button.

- Expected Result: The **Email** button is visible within the Snowflake section.

3. **Click on Email Button**: Click the **Email** button to test its functionality with incomplete data.

- Expected Result: The button is clickable, and an appropriate response is observed, such as a warning message or a handling mechanism for incomplete data.

These test scripts provide a detailed guide for thoroughly verifying the functionality and accessibility of the **Email** button within the Snowflake section, covering positive, negative, and edge scenarios.

Extending test scripts for more complex user stories and tests is a matter of adapting to the multifaceted nature of the requirements. Complex scenarios may require consideration of multiple interactions, dependencies, and potential user paths. To handle this complexity, tests may need to layer different combinations of inputs and account for varying user roles and permissions. Implementing data-driven testing with various datasets can help explore different scenarios. Additionally, structuring the tests in a modular fashion and closely collaborating with different stakeholders can ensure that complex user stories are tested in alignment with both technical specifications and broader business needs. This methodical approach ensures a thorough and accurate validation of the user stories, even as complexity increases.

Let's see how ChatGPT does on a more complex user story.

Prompt	Please write at least one positive test script, at least one negative test script, and at least one edge test script for the user story below. Include a title, an objective, preconditions, assumptions, and detailed step-by-step directions. Include the expected results for each step.
	Epic 3. Email Functionality with Snowflake Data
	User Story 2: Create Flow-Based Wizard to Select a Contact
	• **Goal**: As a Sales Rep, I want a wizard to guide me through the process of selecting a contact to send the email to so that I can choose the correct recipient.
	• **Acceptance Criteria**:
	▪ **Given** a Sales Rep clicks the **Email** button in the Snowflake section,
	▪ **When** the wizard starts,
	▪ **Then** the Sales Rep should be prompted to select a Contact associated with the Account.

Response	**Positive Test Script**
	Title: Verify Flow-Based Wizard for Contact Selection
	Objective: To validate the presence and functionality of the flow-based wizard for selecting a contact after clicking the **Email** button within the Snowflake section.
	Preconditions: Sales Rep is logged in to Salesforce, is on an Account page with a Snowflake section, and has access to at least one Contact associated with the Account.
	Assumptions: The **Email** button, flow-based wizard, and Contact selection are functioning properly.
	Steps:
	1. **Navigate to Snowflake Section**: Ensure the Account page with the Snowflake section has loaded successfully.
	• Expected Result: Account page with Snowflake section is displayed.
	2. **Click on Email Button**: Click the **Email** button within the Snowflake section.
	• Expected Result: Flow-based wizard starts.
	3. **Prompt for Contact Selection**: Observe the wizard prompting to select a Contact.
	• Expected Result: The Sales Rep is prompted to select a Contact associated with the Account.
	4. **Select Contact and Continue**: Select a Contact and proceed through the wizard.
	• Expected Result: Contact is selected, and the wizard continues to the next step as configured.
	Negative Test Script
	Title: Verify Wizard Behavior with No Contacts Associated
	Objective: To ensure that the flow-based wizard handles the scenario where no Contacts are associated with the Account.
	Preconditions: Sales Rep is logged in to Salesforce, is on an Account page with a Snowflake section, and there are no Contacts associated with the Account.
	Assumptions: Proper error handling or informative messaging is implemented for cases without Contacts.

Steps:

1. **Navigate to Snowflake Section**: Ensure the Account page with the Snowflake section has loaded successfully.

- Expected Result: Account page with Snowflake section is displayed.

2. **Click on Email Button**: Click the **Email** button within the Snowflake section.

- Expected Result: Flow-based wizard starts.

3. **Observe Contact Selection Step**: Check the behavior of the wizard without Contacts.

- Expected Result: An informative message or handling mechanism is displayed due to no Contacts being associated with the Account.

Edge Test Script

Title: Verify Wizard Behavior with Large Number of Contacts

Objective: To validate the behavior of the flow-based wizard when there are a large number of Contacts associated with the Account.

Preconditions: Sales Rep is logged in to Salesforce, is on an Account page with a Snowflake section, and there are a large number (e.g., hundreds) of Contacts associated with the Account.

Assumptions: The system can handle large lists of Contacts without performance issues.

Steps:

1. **Navigate to Snowflake Section**: Ensure the Account page with the Snowflake section has loaded successfully.

- Expected Result: Account page with Snowflake section is displayed.

2. **Click on Email Button**: Click the **Email** button within the Snowflake section.

- Expected Result: Flow-based wizard starts.

3. **Prompt for Contact Selection**: Observe the wizard prompting to select a Contact from the large list.

- Expected Result: The Sales Rep is prompted to select a Contact without performance lag or issues.

	4. Select Contact and Continue: Select a Contact from the large list and proceed through the wizard. • Expected Result: Contact is selected, and the wizard continues without performance issues. These test scripts provide detailed guidelines for testing the flow-based wizard functionality, ensuring a smooth user experience for Sales Reps selecting Contacts, even under negative and edge scenarios.

ChatGPT's adaptability to create test scripts for increasingly complex user stories rests in its ability to analyze the multifaceted requirements inherent to sophisticated scenarios. As user stories grow in complexity, involving numerous interactions, dependencies, and conditional paths, the task of manually crafting test scripts can become challenging. ChatGPT, with its understanding of context and ability to process nuanced instructions, can generate coherent and detailed test scripts that align with complex requirements. By taking into account the different layers of interactions and permutations of inputs, it can help in creating structured and comprehensive test scripts. This not only saves time but also ensures that the complex scenarios are accurately represented, allowing developers and testing teams to validate functionality effectively, without being bogged down by the intricacies of manual test creation.

The tests until the preceding edge case have assumed that small amounts of data exist in the system. This is generally a reasonable assumption. At the same time, it is reasonable as part of the test script to explicitly document the data required. ChatGPT can help with this.

Prompt	Please create account test data for Salesforce. Create data for all standard required fields, and name the accounts `Snowflake Email Test No Contacts #1` through `Snowflake Email Test No Contacts #10`. Format the data as CSV.

Response	Certainly! Below is the CSV-formatted data for the 10 Salesforce Accounts. This format can be copied and pasted into a text file with a `.csv` extension for easy import into Salesforce or other applications.
	
	`"Account Name","Type","Industry","Billing Street","Billing City","Billing State","Billing Postal Code","Billing Country","Phone","Website"`
	
	`"Snowflake Email Test No Contacts #1","Customer - Direct","Technology","123 Tech Lane","San Francisco","CA","94105","USA","(123) 456-7890","www.snowflake.com"`
	
	`"Snowflake Email Test No Contacts #2","Customer - Channel","Software","456 Software Street","San Jose","CA","95134","USA","(234) 567-8901","www.snowflake.com"`

```
"Snowflake Email Test No Contacts #3","Customer - Direct",
"Cloud Computing","789 Cloud
Ave","Seattle","WA","98109","USA","(345) 678-9012","www.
snowflake.com"

"Snowflake Email Test No Contacts #4","Customer
- Channel","Technology","123 Tech Lane","San
Francisco","CA","94105","USA","(456) 789-0123","www.
snowflake.com"

"Snowflake Email Test No Contacts #5","Customer
- Direct","Software","456 Software Street","San
Jose","CA","95134","USA","(567) 890-1234","www.
snowflake.com"

"Snowflake Email Test No Contacts #6","Customer -
Channel",
"Cloud Computing","789 Cloud
Ave","Seattle","WA","98109","USA","(678) 901-2345","www.
snowflake.com"

"Snowflake Email Test No Contacts #7","Customer
- Direct","Technology","123 Tech Lane","San
Francisco","CA","94105","USA","(789) 012-3456","www.
snowflake.com"

"Snowflake Email Test No Contacts #8","Customer
- Channel","Software","456 Software Street","San
Jose","CA","95134","USA","(890) 123-4567","www.snowflake.
com"

"Snowflake Email Test No Contacts #9","Customer - Direct",
"Cloud Computing","789 Cloud
Ave","Seattle","WA","98109","USA","(901) 234-5678","www.
snowflake.com"

"Snowflake Email Test No Contacts #10","Prospect","Cloud
Computing","789 Cloud Ave","Seattle","WA","98109","USA","
(012) 345-6789","www.snowflake.com"
```

This CSV format is appropriate for import into Salesforce or any other system that accepts CSV files. Make sure to validate the data with your specific Salesforce instance as there may be custom fields or validation rules that need to be considered.

Similarly, ChatGPT can be used to create account test data for accounts with one or a small number of contacts as well as accounts with hundreds of contact records. Contact data can also be created by ChatGPT. The edge script with hundreds of contacts will be a little more work as there are limits on just how much data ChatGPT will create.

Prompt	Please create contact test data for Salesforce. Create data for all standard required fields, and set the last name of the contacts to `Snowflake Email #1` through `Snowflake Email #200`. Format the data as CSV.

Response	Certainly! Below is a CSV-formatted representation of 200 Salesforce Contacts. Given the large number of records, I'll present a template of the structure and include the details for the first few contacts as examples. You can replicate the pattern to create a full set of 200 contacts.

```
"Last Name","First Name","Account
Name","Title","Email","Phone","Mailing Street","Mailing
City","Mailing State","Mailing Postal Code","Mailing
Country"
"Snowflake Email #1","John","Snowflake Email Test No
Contacts #1","Manager","john1@example.com","(123) 456-
7890","123 Tech Lane","San Francisco","CA","94105","USA"
"Snowflake Email #2","Jane","Snowflake Email Test No
Contacts #1","Engineer","jane2@example.com","(234) 567-
8901","456 Software Street","San Jose","CA","95134","USA"
"Snowflake Email #3","James","Snowflake Email Test No
Contacts #2","CTO","james3@example.com","(345) 678-
9012","789 Cloud Ave","Seattle","WA","98109","USA"
"Snowflake Email #4","Jill","Snowflake Email Test No
Contacts #3","Developer","jill4@example.com","(456) 789-
0123","123 Tech Lane","San Francisco","CA","94105","USA"
...
"Snowflake Email #200","Jasmine","Snowflake Email Test No
Contacts #10","Director","jasmine200@example.com","(901)
234-5678","789 Cloud Ave","Seattle","WA","98109","USA"
```

This structure should align with the standard fields for contacts in Salesforce. Be sure to adapt the `Account Name` field to match the Account names you're associating with each Contact, and feel free to modify other fields as needed to fit your testing requirements.

You can replicate the given lines, changing the appropriate fields, to create the rest of the contacts up to `Snowflake Email #200`. Make sure to align the contact data with the corresponding account data you created earlier.

To create 200 contacts for the account records used for the "large number of contacts" edge script, copy the contacts created by ChatGPT and duplicate them. Alternatively, ChatGPT can be used to create 10 contact records, and then the ChatGPT **Regenerate** button can be clicked 19 times to create the 200 contact records.

If there are additional standard custom fields that need test data created, the prompt can be customized to explicitly name the additional fields.

Prompt	Please create contact test data for Salesforce. Set the last name of the contacts to Snowflake Email Test #1 through Snowflake Email Test #10. Format the data as CSV. Create data for all standard required fields as well as these custom fields: * "VIP Customer", boolean * "LIfetime Value", picklist, values, HIGH, MED, LOW, TBD
Response	Certainly! Here's an example of the contact test data for Salesforce with the given standard required fields and custom fields in CSV format: ``` FirstName,LastName,Email,Phone,VIP Customer,Lifetime Value John,Snowflake Email Test #1,john@test.com,123-456-7890,true,HIGH Jane,Snowflake Email Test #2,jane@test.com,123-456-7891,false,MED Alice,Snowflake Email Test #3,alice@test.com,123-456-7892,false,LOW Bob,Snowflake Email Test #4,bob@test.com,123-456-7893,true,TBD Charlie,Snowflake Email Test #5,charlie@test.com,123-456-7894,false,HIGH Diana,Snowflake Email Test #6,diana@test.com,123-456-7895,true,MED Eva,Snowflake Email Test #7,eva@test.com,123-456-7896,false,LOW Frank,Snowflake Email Test #8,frank@test.com,123-456-7897,true,TBD Grace,Snowflake Email Test #9,grace@test.com,123-456-7898,false,HIGH Henry,Snowflake Email Test #10,henry@test.com,123-456-7899,true,MED ``` This dataset includes the first name, last name (specified as "Snowflake Email Test #n"), email, phone number, VIP Customer status as a boolean (true/false), and Lifetime Value as a picklist with the given values (HIGH, MED, LOW, TBD). Please ensure that these custom fields are properly configured in your Salesforce instance before importing the data.

Determining the number of steps in a test script is a nuanced process that depends on several factors, such as the complexity of the user story and the experience level of the testing team with the technology and application under development. Simple user stories, such as adding a field to a Salesforce page layout, might require only two or three steps, especially if handled by a testing team well versed in Salesforce.

Conversely, complex user stories or working with a team new to Salesforce testing might necessitate dozens of steps. Understanding the user story's intricate details, dependencies, possible interactions, and the experience level of the team guides the process of defining the appropriate number of steps.

However, balancing complexity with clarity is essential. As per testing best practices, limiting the number of steps in a test script helps maintain focus and readability. Although there is no strict rule for the maximum number of steps, keeping a test script within a manageable range of steps (often around 15-20) is generally recommended. This ensures that the test remains comprehensible and maintainable while adequately covering the required functionality. Too many steps can lead to confusion and hinder the efficiency of the testing process, while too few steps might overlook essential details, undermining the test's effectiveness. Therefore, aligning the number of steps with the complexity and the team's proficiency provides a pragmatic approach to crafting precise and effective test scripts.

When desired, the ChatGPT prompt can set the expectation for the number of test steps.

| Prompt | Please write at least one positive test script, at least one negative test script, and at least one edge test script for the user story below. Include a title, an objective, preconditions, assumptions, and step-by-step directions. Provide granular step-by-step directions, 10 or more, for each test. Include the expected results for each step.

Epic 4. **Feedback Mechanism for Incorrect Snowflake Information**

User Story 3: **Create Apex Call Out to Send Feedback to Snowflake**

• **Goal**: As a Sales Rep, I want the feedback about incorrect Snowflake information to be sent to Snowflake so that the necessary corrections can be made.

• **Acceptance Criteria**:

· **Given** a Sales Rep has entered feedback through the Flow-based wizard,

· **When** the feedback is submitted,

· **Then** an Apex Call Out should send the feedback details to Snowflake for review. |
| --- | --- |

Response	Certainly! Below are the positive, negative, and edge test scripts for the given user story.
	Positive Test Script
	Title: Test Apex Call Out Sending Feedback to Snowflake
	Objective: Verify that the Apex Call Out sends feedback to Snowflake correctly.
	Preconditions:
	• Sales Rep is logged in to Salesforce.
	• The feedback flow-based wizard is configured properly.
	Assumptions:
	• Snowflake integration is configured correctly.
	Step-by-Step Directions:
	1. Navigate to the Snowflake section on the Account page.
	• Expected Result: Snowflake section is visible.
	2. Click on the **Feedback** button.
	• Expected Result: The flow-based feedback wizard starts.
	3. Enter valid feedback about incorrect Snowflake information.
	• Expected Result: Feedback is accepted.
	4. Click on the **Submit** button within the wizard.
	• Expected Result: The feedback is submitted and the Apex Call Out is initiated.
	5. Verify that the Apex Call Out sends the feedback to Snowflake.
	• Expected Result: The correct feedback information is sent to Snowflake.
	6. Check the acknowledgment received from Snowflake.
	• Expected Result: Snowflake acknowledges the receipt of feedback.
	7. Validate that the feedback details are logged within Salesforce.
	• Expected Result: The log in Salesforce confirms that the feedback was sent successfully.

8. Verify that the feedback is stored in Snowflake correctly.

- Expected Result: Feedback is found in Snowflake's database with correct details.

9. Repeat steps 3-8 with different valid feedback entries.

- Expected Result: All different valid feedback entries are handled correctly.

10. Log out from Salesforce.

- Expected Result: Successful logout.

Negative Test Script

Title: Test Invalid Feedback Submission through Apex Call Out

Objective: Ensure that invalid feedback does not get sent to Snowflake.

Preconditions:

- Sales Rep is logged in to Salesforce.
- The feedback flow-based wizard is configured properly.

Assumptions:

- Snowflake integration is configured correctly.

Step-by-Step Directions:

1. Navigate to the Snowflake section on the Account page.

- Expected Result: Snowflake section is visible.

2. Click on the **Feedback** button.

- Expected Result: The flow-based feedback wizard starts.

3. Enter invalid feedback (e.g., empty, special characters).

- Expected Result: Error message is displayed.

4. Attempt to click on the **Submit** button within the wizard.

- Expected Result: Submission is not allowed; error message remains.

5. Correct the feedback and submit.

- Expected Result: Feedback is accepted and the Apex Call Out is initiated.

6. Disconnect Snowflake integration.

- Expected Result: Snowflake integration is disconnected.

7. Repeat steps 3-5.

- Expected Result: Error message indicating that Snowflake is not reachable.

8. Reconnect Snowflake integration.

- Expected Result: Snowflake integration is reconnected successfully.

9. Repeat steps 3-5 with valid feedback.

- Expected Result: Feedback is sent successfully to Snowflake.

10. Log out from Salesforce.

- Expected Result: Successful logout.

Edge Test Script

Title: Test Apex Call Out with Boundary Conditions

Objective: Verify how the Apex Call Out handles feedback at its boundary conditions.

Preconditions:

- Sales Rep is logged in to Salesforce.
- The feedback flow-based wizard is configured properly.

Assumptions:

- Snowflake integration is configured correctly.

Step-by-Step Directions:

1. Navigate to the Snowflake section on the Account page.

- Expected Result: Snowflake section is visible.

2. Click on the **Feedback** button.

- Expected Result: The flow-based feedback wizard starts.

3. Enter feedback with exactly the maximum allowable characters.

- Expected Result: Feedback is accepted.

4. Click on the **Submit** button within the wizard.

- Expected Result: The feedback is submitted and the Apex Call Out is initiated.

5. Verify that the feedback is sent to Snowflake.

- Expected Result: Feedback is sent to Snowflake.

6. Enter feedback with exactly one character above the maximum allowable characters.

- Expected Result: Error message indicating that the feedback exceeds the allowable characters.

7. Enter feedback with exactly the minimum allowable characters.

- Expected Result: Feedback is accepted.

8. Click on the **Submit** button within the wizard.

- Expected Result: The feedback is submitted and the Apex Call Out is initiated.

9. Verify that the feedback is sent to Snowflake.

- Expected Result: Feedback is sent to Snowflake.

10. Enter feedback with exactly one character below the minimum allowable characters.

- Expected Result: Error message indicating that the feedback is below the allowable characters.

These tests cover the standard positive, negative, and edge cases for the Apex Call Out to send feedback to Snowflake, ensuring the functionality aligns with the defined goal and acceptance criteria of the user story.

The use of spreadsheets in capturing and running tests is a common and flexible approach, especially in environments where a dedicated commercial testing tool may not be employed. Spreadsheets allow teams to organize, format, and execute test scripts methodically, providing an accessible and low-cost solution. They can be easily shared and updated, and often include features that can assist in the tracking and analysis of test execution results.

However, the rise of commercial testing tools has expanded the horizons of software testing, offering powerful functionalities tailored to testing requirements. These tools can automate several tasks, provide detailed reporting, and foster collaboration among teams. They often enable integration with various platforms, including Salesforce, and support importing test scripts from different formats, including spreadsheets.

ChatGPT aligns with both these approaches by offering the ability to write test scripts that can be seamlessly loaded into spreadsheets or testing tools. This adaptability ensures that the testing process remains efficient and scalable, whether using simple spreadsheet-based management or more complex commercial testing tools. By understanding the particular structure and requirements of the given tools, ChatGPT facilitates the rapid creation of detailed and accurate test scripts, bridging the gap between manual efforts and automated tool capabilities, thus serving a broader range of testing scenarios.

Prompt	Please rewrite the test scripts for Epic 4 User Story 3 to be a CSV file that can be loaded into a spreadsheet or testing tool. Include the Epic and User Story names in the CSV file.
Response	Certainly! Below are the CSV representations for the positive, negative, and edge test scripts for Epic 4, User Story 3. I have included the Epic and User Story names in the CSV file, making it suitable for import into a spreadsheet or testing tool.

Positive Test Script CSV

```
"Epic","User
Story","Type","Title","Step","Action","Expected Result"
"Epic 4. Feedback Mechanism for Incorrect Snowflake
Information","User Story 3: Create Apex Call Out to
Send Feedback to Snowflake","Positive","Test Apex Call
Out Sending Feedback to Snowflake","1","Navigate to the
Snowflake section on the Account page.","Snowflake section
is visible."
"Epic 4. Feedback Mechanism for Incorrect Snowflake
Information","User Story 3: Create Apex Call Out to Send
Feedback to Snowflake","Positive","Test Apex Call Out
Sending Feedback to Snowflake","2","Click on the 'Feedback'
button.","The Flow-based feedback wizard starts."
...[repeat for all positive test steps]...
```

Negative Test Script CSV

```
"Epic","User
Story","Type","Title","Step","Action","Expected Result"
"Epic 4. Feedback Mechanism for Incorrect Snowflake
Information","User Story 3: Create Apex Call Out to Send
Feedback to Snowflake","Negative","Test Invalid Feedback
Submission through Apex Call Out","1","Navigate to the
Snowflake section on the Account page.","Snowflake section
is visible."
"Epic 4. Feedback Mechanism for Incorrect Snowflake
Information","User Story 3: Create Apex Call Out to
Send Feedback to Snowflake","Negative","Test Invalid
Feedback Submission through Apex Call Out","2","Click on
the 'Feedback' button.","The Flow-based feedback wizard
starts."
...[repeat for all negative test steps]...
```

> **Edge Test Script CSV**
>
> ```
> "Epic","User
> Story","Type","Title","Step","Action","Expected Result"
> ```
> ```
> "Epic 4. Feedback Mechanism for Incorrect Snowflake
> Information","User Story 3: Create Apex Call Out to Send
> Feedback to Snowflake","Edge","Test Apex Call Out with
> Boundary Conditions","1","Navigate to the Snowflake section
> on the Account page.","Snowflake section is visible."
> ```
> ```
> "Epic 4. Feedback Mechanism for Incorrect Snowflake
> Information","User Story 3: Create Apex Call Out to
> Send Feedback to Snowflake","Edge","Test Apex Call Out
> with Boundary Conditions","2","Click on the 'Feedback'
> button.","The Flow-based feedback wizard starts."
> ```
> ```
> ...[repeat for all edge test steps]...
> ```
>
> These CSV files include the Epic and User Story names, alongside the structured representation of the test scripts. This format allows for clear mapping of each test to the corresponding Epic and User Story and ensures that the test details are easily accessible for analysis and execution.

Please note that ChatGPT did not produce the entirety of each CSV file. After the first handful of rows, it said: "[repeat for all edge test steps]". This can be addressed with a prompt for each individual CSV file:

Please rewrite the positive test script for Epic 4 User Story 3 to be a CSV file that can be loaded into a spreadsheet or testing tool. Include the Epic and User Story names in the CSV file. Include all steps.

Working with ChatGPT is an iterative process. Create refining prompts until the responses stop improving and only then start manually updating responses.

Troubleshooting testing issues

Salesforce test scripts presuppose a certain amount of knowledge about the Salesforce UI and Salesforce overall. As Salesforce includes more features and the features have gotten more complex, this has gotten more challenging for testers. Another place ChatGPT can help with testing is to provide insight on the step-by-step directions.

Prompt	The first step in a positive test case is below. Please provide directions on how to navigate from the standard Salesforce landing page to the Account page in order to complete this first step.
	Epic 4. **Feedback Mechanism for Incorrect Snowflake Information**
	User Story 3: **Create Apex Call Out to Send Feedback to Snowflake**
	Step-by-Step Directions:
	1. Navigate to the Snowflake section on the Account page.
	• Expected Result: Snowflake section is visible.
Response	Certainly! The first step in this positive test case can be achieved by navigating from the standard Salesforce landing page to the Account page and finally to the Snowflake section. Here are the detailed directions for this step:
	1. Log in to Salesforce using valid credentials.
	2. Once logged in, you'll be taken to the Salesforce home page or landing page.
	3. Click on the App Launcher (grid icon) located at the top-left corner of the page.
	4. In the search bar, type Accounts and select the Accounts object from the drop-down results.
	5. You will be directed to the Accounts list view. Search for the specific Account you wish to view, and click on the Account name to open the Account page.
	6. Once on the Account page, scroll down or look through the tabs to find the Snowflake section, if it's customized in your layout.
	7. Click on the Snowflake section, if it is clickable, to fully expand or access it.
	Expected Result: The Snowflake section is visible on the Account page.
	These directions guide a user from the standard Salesforce landing page, through the application's navigation structure, to the Account page, and finally to the specific Snowflake section that needs to be tested. It considers the typical Salesforce user interface but may need to be tailored to your organization's specific customization and layout.

Summary

In this chapter, readers have gained a comprehensive understanding of the pivotal role of testing in Salesforce development, emphasizing the integration of ChatGPT to enhance this process. The material provided a deep dive into the fundamentals of Salesforce testing, revealing the intricacies and the value of ChatGPT in creating more effective testing strategies. Readers learned how to craft clear, comprehensive, and relevant test cases for Salesforce deployments, with an emphasis on generating test data and leveraging ChatGPT's guidance and expertise. A practical case study illustrated the journey from user stories to Salesforce project test scripts, providing real-world applicability. The skills acquired include a mastery of effective Salesforce testing principles and practices, the ability to craft high-quality test scripts, and proficiency in understanding and optimizing these scripts within the Salesforce framework. This chapter has equipped readers with the knowledge and skills to ensure their Salesforce deployments are robust, reliable, and aligned with user expectations.

In the next chapter, readers will learn about the critical role of debugging in Salesforce deployments and how integrating ChatGPT can significantly enhance this process. The chapter begins by establishing the foundational aspects of Salesforce debugging, highlighting ChatGPT's transformative impact in this area. Readers will learn to quickly identify and diagnose issues in Salesforce deployments using ChatGPT's extensive knowledge and intuitive guidance. The chapter delves into the specifics of tracing and analyzing Salesforce debug logs, demonstrating how to interpret and utilize these logs effectively with ChatGPT's assistance. Additionally, it covers real-time problem resolution, showcasing how ChatGPT can provide immediate recommendations, code snippets, and best practices. The focus then shifts to debugging best practices and the importance of continuous improvement, using ChatGPT to stay updated and refine debugging strategies. By the end of the chapter, readers will have gained skills in understanding Salesforce debugging fundamentals, rapid issue identification, log analysis proficiency, effective problem resolution, and adhering to evolving best practices, all augmented by the capabilities of ChatGPT. This knowledge will empower them to tackle debugging challenges more efficiently and effectively, ensuring their Salesforce deployments are robust and reliable.

13

Using ChatGPT for Salesforce Debugging

This chapter will dive into the world of Salesforce debugging, powered by the synergies between ChatGPT and the debugging process. This chapter reviews the fundamental principles of Salesforce debugging, laying the groundwork for a deep understanding of the subject. It showcases the transformative potential of ChatGPT in this domain, introducing developers to a wealth of advantages and insights that can significantly reduce the time spent on problem detection and resolution.

The chapter explains how to leverage ChatGPT's vast knowledge to streamline the issue identification process. The chapter delves into the complexities of Salesforce debug logs, providing guidance on how to interpret, filter, and analyze them effectively with ChatGPT's assistance. Real-time resolution takes center stage, demonstrating the power of instant problem-solving as ChatGPT provides on-the-spot recommendations, code snippets, and best practices.

This will help establish a strong foundation in debugging principles, skills in rapid issue identification, log analysis proficiency, effective problem resolution, and adherence to industry best practices. ChatGPT acts as a constant guide and resource, ensuring that the debugging process remains efficient, effective, and aligned with evolving industry standards.

Understanding Salesforce debugging and ChatGPT's advantages

ChatGPT possesses an extraordinary capability akin to eidetic memory, allowing it to review and retain vast amounts of information relevant to Salesforce debugging. This remarkable ability enables ChatGPT to serve as an invaluable companion during the debugging process, as it can quickly reference a wealth of Salesforce-related knowledge to determine the next steps in resolving issues. Whether it's identifying anomalies, interpreting debug logs, or suggesting best practices, ChatGPT's memory empowers it to provide precise and contextually relevant guidance, making it an indispensable asset for developers seeking to streamline and enhance their debugging efforts in Salesforce deployments.

Leveraging ChatGPT for debugging in Salesforce projects presents a unique challenge—it's not just about the availability of knowledge but also the precision in applying that knowledge to resolve specific issues. The wealth of information at ChatGPT's disposal is immense, spanning a broad spectrum of Salesforce-related topics. Therefore, the real challenge lies in determining which pieces of knowledge are most relevant and on the shortest path to issue resolution. This means that when developers seek assistance from ChatGPT for debugging, providing more context and information is often better. By presenting a detailed picture of the problem at hand, developers can guide ChatGPT toward the most pertinent insights and solutions.

Additionally, treating ChatGPT as a collaborative partner in the debugging process can be highly effective. Instead of simply requesting information, developers can engage in a dialogue with ChatGPT, asking questions such as, "What additional information may be helpful in diagnosing this issue?" This approach fosters a dynamic interaction where ChatGPT actively participates in problem-solving by suggesting specific data points or aspects of the code that could shed light on the problem. By treating ChatGPT as a knowledgeable collaborator and asking for its input on the most relevant information, developers can navigate the vast sea of knowledge effectively and arrive at solutions faster and with greater precision.

Issue identification with ChatGPT

Real-world scenarios will illuminate the process of using ChatGPT to address common challenges encountered by developers in Salesforce projects. Two examples in this chapter include scenarios where Apex hits a governor limit and where an Apex runtime error is encountered. These hands-on examples will provide insights into how ChatGPT can serve as a knowledgeable partner in identifying, diagnosing, and resolving these issues efficiently, ultimately enhancing the debugging process in Salesforce development.

Governor limits

Salesforce governor limits are a set of resource constraints and limitations imposed by the Salesforce platform to ensure the efficient use of its shared resources and to maintain system stability and performance. They exist to prevent individual customers or applications from monopolizing resources, which could lead to degraded performance for other users on the platform. These limits are in place to ensure that no single operation, user, or application consumes excessive system resources.

Developers often encounter various types of governor limit issues, including hitting the maximum number of SOQL queries or DML statements, exceeding CPU time limits, reaching concurrent request limits, and exceeding the number of records processed in a single transaction. These limits can pose challenges when writing complex code or designing intricate workflows, as they require careful planning and optimization to avoid hitting these constraints. Understanding and effectively managing governor limits are essential skills for Salesforce developers to ensure their applications operate smoothly and efficiently within the Salesforce ecosystem.

While there are many governor limits and related error messages, one limit typically encountered as usage of Salesforce increases is the limit of 100 SOQL queries in an Apex transaction:

System.LimitException: Too many SOQL queries: 101

Let's start by asking ChatGPT for more information as well as for some reference material:

Prompt	Please provide a detailed explanation of what causes this Salesforce error: "System. LimitException: Too many SOQL queries: 101" as well as some reference material so that I can learn more about this error.
Response	The Salesforce error message "System.LimitException: Too many SOQL queries: 101" occurs when a Salesforce transaction, such as an Apex trigger or a Visualforce page, exceeds the governor limit for the number of SOQL (Salesforce Object Query Language) queries it can perform in a single transaction. In this specific case, the error indicates that the code executed a total of 101 SOQL queries in a single transaction, surpassing the established limit. Here's a breakdown of the key elements related to this error: 1. SOQL Queries: SOQL is Salesforce's query language used to retrieve data from the Salesforce database. Each query to the database consumes a certain amount of resources. 2. Governor Limits: Salesforce imposes governor limits to ensure fair usage of system resources on its shared platform. One of these limits is the maximum number of SOQL queries allowed per transaction. 3. Transaction: In Salesforce, a transaction refers to a sequence of operations that are treated as a single unit of work. This can include, but is not limited to, executing Apex code, processing triggers, and performing DML (Data Manipulation Language) operations such as inserts, updates, and deletes. To resolve this error, developers need to review and optimize their code to reduce the number of SOQL queries executed within a single transaction. Techniques such as bulkifying queries (performing operations on multiple records in a single query), using collections to aggregate data, and caching results can help minimize the number of queries. For more information and detailed guidance on dealing with this error and best practices for working with governor limits in Salesforce, I recommend referring to the official Salesforce documentation on governor limits: - [Salesforce Governor Limits](`https://developer.salesforce.com/docs/atlas.en-us.apexcode.meta/apexcode/apex_gov_limits.htm`) This documentation provides in-depth explanations of governor limits, including limits on SOQL queries, along with examples and strategies for working within these limits effectively.

This is helpful but doesn't pinpoint the problem in the existing code. Let's ask ChatGPT for help in figuring out the basis for this error:

Prompt	What, specifically, should I look for in the Apex code to isolate what is causing this error?

Response	To isolate and identify the specific part of your Apex code that is causing the "Too many SOQL queries" error in Salesforce, you should perform a systematic review of your code base. Here are steps and things to look for:
	
	1. Identify the Error Location:
	
	   - First, identify the specific transaction or operation that triggered the error. This could be a trigger, Visualforce page, or other Apex code.
	
	2. Review SOQL Query Usage:
	
	   - Examine all SOQL queries in the relevant code. Pay close attention to the number and placement of queries within loops.
	
	   - Look for queries within loops: If you find SOQL queries within a loop (e.g., a for loop that queries inside it), these are prime candidates for optimization.
	
	   - Review the total number of queries executed in the transaction. Consider whether any nested queries or subqueries are contributing to the count.
	
	3. Optimize SOQL Queries:
	
	   - Refactor queries to reduce the total number of queries made. Consider using aggregate queries, subqueries, and bulk queries to retrieve data efficiently.
	
	   - Avoid unnecessary queries: Ensure that queries are only made when needed and that you're not querying the same data multiple times in the same transaction.
	
	   - Leverage collections: Use collections (lists, sets, maps) to store and manipulate data rather than querying repeatedly.
	
	4. Implement Caching:
	
	   - Consider implementing caching mechanisms, such as the Platform Cache or custom caching strategies, to store frequently used data to reduce the need for repetitive queries.
	
	5. Bulkify Triggers:
	
	   - If the error is occurring within a trigger, ensure that your trigger logic is bulkified. This means that it can handle multiple records in a single transaction efficiently.

6. Testing and Debugging:

- Use Salesforce debugging tools, such as debug logs and the Developer Console, to trace the execution of your code and identify which queries are being executed and how many times.

- Utilize system.debug statements strategically to log relevant information about the execution flow.

7. Code Reviews and Best Practices:

- Collaborate with colleagues or team members to review your code for best practices and optimization opportunities.

8. Unit Testing:

- Write comprehensive unit tests that cover various scenarios, including scenarios that could potentially hit governor limits. This helps you catch issues early in development.

9. Consider Asynchronous Processing:

- In some cases, you may need to consider offloading heavy processing or queries to asynchronous jobs, such as Batch Apex or Queueable Apex, to avoid hitting limits.

By following these steps and best practices, you can systematically identify and address the root cause of the "Too many SOQL queries" error in your Apex code, ensuring that your code is optimized for efficient data retrieval and processing in Salesforce.

Based on the preceding response from ChatGPT, the following Apex was identified as the issue. There is an SQL call to get contacts inside the `for` loop that processes account records:

```
public class AccountProcessor {
    public static void updateContactsForAccounts(List<Account>
accounts) {
        List<Contact> contactsToUpdate = new List<Contact>();

        for (Account acc : accounts) {
            List<Contact> relatedContacts = [SELECT Id, FirstName,
LastName FROM Contact WHERE AccountId = :acc.Id];

            // Loop through the contacts and update them
            for (Contact con : relatedContacts) {
                con.FirstName = 'UpdatedFirstName';
                con.LastName = 'UpdatedLastName';
                contactsToUpdate.add(con);
            }
        }
```

```
            // Update the modified contacts
        if (!contactsToUpdate.isEmpty()) {
            update contactsToUpdate;
        }
    }
}
```

Let's ask ChatGPT to rewrite the Apex to address the issue by moving the Contact SOQL query outside of the Account `for` loop:

Prompt	Please rewrite this Apex code to move the Contact SOQL query outside of the for loop that processes Account records in order to address the "System.LimitException: Too many SOQL queries: 101" error.
	<pre>public class AccountProcessor {
    public static void
updateContactsForAccounts(List<Account> accounts) {
        List<Contact> contactsToUpdate = new
List<Contact>();
        for (Account acc : accounts) {
            List<Contact> relatedContacts = [SELECT Id,
FirstName, LastName FROM Contact WHERE AccountId = :acc.
Id];
                // Loop through the contacts and update them
            for (Contact con : relatedContacts) {
                    con.FirstName = 'UpdatedFirstName';
                    con.LastName = 'UpdatedLastName';
                    contactsToUpdate.add(con);
            }
        }
        // Update the modified contacts
        if (!contactsToUpdate.isEmpty()) {
            update contactsToUpdate;
        }
    }
}</pre>

Response	To address the "System.LimitException: Too many SOQL queries" error and improve the code's performance, you can move the Contact query outside of the loop that processes Account records. Here's the modified code:

```
public class AccountProcessor {
    public static void
updateContactsForAccounts(List<Account> accounts) {
        List<Contact> contactsToUpdate = new
List<Contact>();
        // Collect Account Ids for all accounts in the
input list
        Set<Id> accountIds = new Set<Id>();
        for (Account acc : accounts) {
            accountIds.add(acc.Id);
        }
        // Query all related contacts for the collected
Account Ids in a single query
        List<Contact> relatedContacts = [SELECT Id,
FirstName, LastName, AccountId FROM Contact WHERE AccountId
IN :accountIds];
        // Map contacts by Account Id for easy reference
        Map<Id, List<Contact>> accountToContactsMap = new
Map<Id, List<Contact>>();
        for (Contact con : relatedContacts) {
            if (!accountToContactsMap.containsKey(con.
AccountId)) {
                accountToContactsMap.put(con.AccountId, new
List<Contact>());
            }
            accountToContactsMap.get(con.AccountId).
add(con);
        }
        // Loop through the accounts and their related
contacts
        for (Account acc : accounts) {
            if (accountToContactsMap.containsKey(acc.Id)) {
                List<Contact> relatedContactsForAccount =
accountToContactsMap.get(acc.Id);
                for (Contact con :
relatedContactsForAccount) {
                    con.FirstName = 'UpdatedFirstName';
                    con.LastName = 'UpdatedLastName';
                    contactsToUpdate.add(con);
                }
            }
        }
```

```
                    // Update the modified contacts
                    if (!contactsToUpdate.isEmpty()) {
                        update contactsToUpdate;
                    }
                }
            }
```

In this modified code, we first collect the Account Ids from the input list of accounts and then perform a single SOQL query to retrieve all related contacts for those Account Ids. The retrieved contacts are then organized into a map for easy reference. This approach significantly reduces the number of SOQL queries executed, addressing the governor limit issue.

With ChatGPT's assistance, in a matter of minutes, a Salesforce governor limit error has been addressed. When faced with an Apex error such as "*System.LimitException: Too many SOQL queries*," ChatGPT can quickly provide a concise explanation of the root cause, shedding light on why the error occurred. Moreover, it offers valuable guidance by suggesting resources and references, allowing the affected developer to delve deeper into the topic at their own pace. Whether a novice or an experienced developer, ChatGPT's ability to rapidly provide insights and direction can be a game-changer in troubleshooting and learning from errors.

Additionally, ChatGPT can swiftly rewrite the Apex code to address the specific error. In just a few minutes, ChatGPT modified the code snippet to not only resolve the issue but also follow best practices, ensuring the code remains efficient and scalable. This collaborative approach, where ChatGPT acts as a knowledgeable partner, accelerates the development process and fosters a culture of continuous learning and improvement in Salesforce development.

Apex runtime errors

Salesforce Apex runtime errors are issues that occur while executing Apex code during the runtime or execution phase of a Salesforce transaction. Unlike compile-time errors, which are detected during code compilation, runtime errors manifest when the code is actually running. These errors can disrupt the normal flow of a Salesforce transaction and may result in exceptions or unexpected behavior.

One common example of a runtime error in Salesforce is the "*NullPointerException*" or "*Attempt to de-reference a null object*" error. This occurs when the code attempts to access or manipulate an object or variable that doesn't have a valid value assigned to it, typically because it's null. Salesforce developers must be vigilant in handling such runtime errors by implementing error-checking mechanisms, such as conditional statements and `try-catch` blocks, to gracefully handle unexpected situations and prevent transaction failures. Understanding and effectively addressing runtime errors is a crucial skill for Apex developers to ensure the reliability and robustness of their Salesforce applications.

An example of what a NullPointerException might look like is provided next. Please note this is an example created for this explanation, which is why the timestamp and line numbers are placeholders:

```
08:45:30.123 (123456789)|EXCEPTION_THROWN|[4]|System.
NullPointerException: Attempt to de-reference a null object
08:45:30.123 (123456790)|FATAL_ERROR|System.NullPointerException:
Attempt to de-reference a null object
```

Let's ask ChatGPT what to look for in the Apex code that might cause this error:

Prompt	What do I look for in the Salesforce Apex code to identify what is causing this error?
	08:45:30.123 (123456789)|EXCEPTION_THROWN|[4]|System.NullPointerException: Attempt to de-reference a null object
	08:45:30.123 (123456790)|FATAL_ERROR|System.NullPointerException: Attempt to de-reference a null object
Response	To identify what is causing a "System.NullPointerException" error in Salesforce Apex code, you should focus on the following key information from the debug log:
	1. Line Number: The line number where the exception was thrown is indicated in square brackets `[4]` in your example. This line number corresponds to the specific line of code in your Apex class or trigger where the error occurred.
	2. Error Message: The error message itself provides valuable information. In this case, it says, "Attempt to de-reference a null object." This suggests that the error is related to attempting to access or manipulate a null object.
	Once you have this information, follow these steps to identify the cause of the error:
	1. Go to the indicated line number in your Apex code (in this case, line 4). This is the location where the error occurred.
	2. Review the code on that line and the surrounding lines to identify any variables, objects, or references that could be null at that point in the code.
	3. Check if there are any assignments or operations that might result in a null value for an object or variable.
	4. Pay special attention to any code that involves accessing object properties, calling methods, or performing operations on variables, as these are common sources of "NullPointerException."
	5. Implement error-checking mechanisms such as conditional statements (`if` checks) or try-catch blocks to handle null values and prevent the error from occurring.

> By carefully examining the code around the line where the error was thrown and considering the context, you can pinpoint the root cause of the "System. NullPointerException" and make the necessary adjustments to handle null values appropriately and prevent the error from happening in the future.

The code throwing the error is shown next. This is an example written specifically to throw this error; the actual code will be more complicated and take more work to trace to the root cause of the null exception, but the process is still the same!

```
public class NullPointerExceptionExample {
    public static void main() {
        // Attempting to access a method of a null object
        String nullString = null;
        Integer stringLength = nullString.length();
    }
}
```

In this example, a nullString variable is declared and assigned a null value, which means it does not reference any valid String object. Then, the code tries to access the length() method of nullString. Since nullString is null and doesn't refer to any String object, this line of code will result in a "NullPointerException."

To avoid such errors in Salesforce Apex, developers need to ensure that variables and objects are properly initialized before attempting to access their methods or properties. Handling null values with conditional statements and try-catch blocks is a best practice to prevent such exceptions and maintain the stability of the application.

Let's ask ChatGPT to rewrite the Apex to preclude the error:

Prompt	Please rewrite this Salesforce Apex code to preclude it from throwing the Null PointerException error. Additionally, add an if statement to check to make sure the nullString value is not null before performing operations on it and add a try-catch-finally block to preclude this code from terminating unexpectedly.
| | ```
public class NullPointerExceptionExample {
    public static void main() {
        // Attempting to access a method of a null object
        String nullString = null;
        Integer stringLength = nullString.length();
    }
}
``` |
```

Response	Certainly, here's the Salesforce Apex code with error handling to preclude the "NullPointerException" error:

```
public class NullPointerExceptionExample {
    public static void main() {
        // Attempting to access a method of a null object
        String nullString = null;
        Integer stringLength = null;
        try {
            // Check if nullString is not null before
accessing its length
            if (nullString != null) {
                stringLength = nullString.length();
            } else {
                // Handle the case when nullString is null
                // You can assign a default value or take
appropriate action
                stringLength = 0; // Assigning a default
value of 0
            }
        } catch (Exception e) {
            // Handle any exceptions that may occur during
the code execution
            System.debug('An error occurred: ' +
e.getMessage());
        } finally {
            // Code to be executed regardless of whether
an exception occurred or not
            System.debug('The string length is: ' +
stringLength);
        }
    }
}
```

> In this modified code:
>
> 1. We added a `try` block to encapsulate the code that might throw an exception.
>
> 2. Inside the `try` block, we added an `if` statement to check if `nullString` is not null before attempting to access its length. If it's null, we assign a default value of 0 to `stringLength` to avoid the "NullPointerException."
>
> 3. We added a `catch` block to handle any exceptions that may occur during the execution of the code. It will capture and log the exception message if one occurs.
>
> 4. Finally, we included a `finally` block to specify code that will run regardless of whether an exception occurred or not. In this case, it logs the value of `stringLength`.
>
> With these modifications, the code will handle null values gracefully and prevent the "NullPointerException" from terminating the program unexpectedly.

With ChatGPT's assistance, the process of identifying and resolving a "NullPointerException" error in the Apex code took only a matter of minutes. ChatGPT provided clear guidance on how to handle `null` values, introduced the concept of error handling with `try-catch` blocks, and helped update the code to avoid unexpected terminations. It's worth noting that while the immediate issue was addressed efficiently, the experience also highlighted the importance of having a comprehensive Salesforce org-wide error-handling framework in place. Such a framework can help standardize error-handling practices and enhance the overall robustness of an organization's Apex code base. However, the details of implementing such a framework go beyond the scope of this section of the book, emphasizing the need for ongoing best practices and development standards in Salesforce projects.

Tracing and analyzing debug logs

Salesforce debug logs are a critical tool for developers and administrators to gain insights into the inner workings of their Salesforce applications. These logs provide a detailed record of the execution of code, including Apex classes, triggers, and other components. Debug logs allow users to trace the flow of their code, monitor variable values, and identify any issues or errors that may arise during execution.

To utilize Salesforce debug logs effectively, developers can set up log levels to specify the level of detail they want to capture. These log levels can be customized to focus on specific components or operations within their code, ensuring that the log remains manageable in size while still providing the necessary information for debugging. Debug logs can be generated for specific users, so developers can analyze the execution of code within the context of a particular user's actions.

The following section delves into how ChatGPT can be employed to analyze Salesforce debug logs. With ChatGPT's assistance, users can learn how to interpret log entries, identify potential issues, and even receive guidance on how to resolve problems effectively. This collaborative approach to debugging, facilitated by ChatGPT, can significantly streamline the development and troubleshooting process in Salesforce projects, leading to more robust and reliable applications.

A sample debug log snippet is shown next. This is a simplified snippet created for this example:

```
26.0.1 (123456789)|CODE_UNIT_
STARTED|[EXTERNAL]|01q7E000005LW7b|MyApexClass.myMethod
26.0.1 (123456790)|METHOD_ENTRY|[5]|01p7E0000012ABc|MyClass.
constructor
26.0.1 (123456791)|VARIABLE_ASSIGNMENT|[8]|myVariable|null
26.0.1 (123456792)|METHOD_EXIT|[5]|MyClass.constructor
26.0.1 (123456793)|METHOD_ENTRY|[10]|01q7E000005LW7b|MyApexClass.
anotherMethod
26.0.1 (123456794)|VARIABLE_ASSIGNMENT|[12]|myString|Hello, World!
26.0.1 (123456795)|SYSTEM_METHOD_ENTRY|[15]|System.debug(ApexLog): My
debug message
26.0.1 (123456796)|SYSTEM_METHOD_EXIT|[15]|System.debug(ApexLog): My
debug message
26.0.1 (123456797)|METHOD_EXIT|[10]|MyApexClass.anotherMethod
```

Let's dive into getting help from ChatGPT to suss out the meaning of the rows in this debug log snippet:

Prompt	The line below is from a Salesforce debug log. What does it mean, and how should I interpret it appearing in the debug log? 26.0.1 (123456789)|CODE_UNIT_STARTED|[EXTERNAL]|01q7E000005LW7b|MyApexClass.myMethod
Response	about the start of a code unit execution. Here's a breakdown of its components: - `26.0.1 (123456789)`: This part includes a timestamp indicating the time when the log entry was generated. It also typically includes information about the duration of the execution. The specific numbers and format may vary depending on your Salesforce instance and log configuration. - `CODE_UNIT_STARTED`: This is a log event type that signifies the start of a code unit execution. Code units can include triggers, Apex classes, and other pieces of code. - `[EXTERNAL]`: This part of the entry can indicate whether the code unit is internal (related to your organization's custom code) or external (related to managed packages or Salesforce platform code). - `01q7E000005LW7b`: This is a unique identifier for the code unit being executed. It can be used to associate the log entry with a specific part of your Salesforce code base, such as an Apex class or trigger. The line you provided from a Salesforce debug log is a log entry that provides information - `MyApexClass.myMethod`: This is the name of the specific code unit that is starting its execution. In this case, it's an Apex class named `MyApexClass` and a method or function named `myMethod`.

	Interpreting this log entry, it tells you that at the specified timestamp, a code unit named `MyApexClass.myMethod` is beginning to execute. Debug logs can include these entries for various code units, allowing developers to trace the flow of their code and understand the order in which different components are invoked during a particular operation or transaction. This information is valuable for debugging and performance analysis.

This explanation is helpful in getting started on understanding the debug log row. One powerful capability of ChatGPT is the ability to drill down into more granular levels of detail. Let's explore the meaning of internal and external in the context of code:

Prompt	What does it mean when a row from a Salesforce debug log says the code is "EXTERNAL"? If the code is throwing an error, can I fix it myself, or am I required to follow up with Salesforce and/or the vendor of the managed package containing the code?
Response	In a Salesforce debug log, when a row specifies that the code is "EXTERNAL," it means that the code being executed is part of a managed package or a component provided by Salesforce, rather than being custom code developed by your organization. Here's what it implies:
	1. Managed Package or Salesforce Platform Code: "EXTERNAL" code indicates that the specific code unit or operation being logged is associated with a managed package or the core Salesforce platform itself. Managed packages are typically developed by third-party vendors and are designed to provide additional functionality or features to Salesforce organizations.
	2. Limited Customization: Since this code is not part of your organization's custom code base, you have limited control over it. You can't directly modify or fix it yourself unless the vendor of the managed package provides customization options.
	3. Responsibility for Errors: If the "EXTERNAL" code is throwing an error or behaving unexpectedly, your first step should be to consult the documentation and support resources provided by the managed package vendor. They may have guidance on how to configure or use their package correctly. If you believe it's a bug or issue with the managed package itself, you should report it to the vendor's support or development team.
	4. Salesforce Support: In some cases, if the issue is related to the Salesforce platform itself and not a specific managed package, you may need to contact Salesforce Support for assistance.
	In summary, when encountering "EXTERNAL" code in a debug log, you should seek support and guidance from the relevant package vendor or Salesforce Support. You typically won't be able to directly fix or modify this code yourself unless there are customization options provided by the vendor.

It is possible to keep drilling down into the details, as well as ask for reference material/additional reading material. Next, let's explore a system method entry row:

Prompt	What does this row in a Salesforce debug log mean? What is it appearing in the log? Do I need to do anything about it?
	26.0.1 (123456795)|SYSTEM_METHOD_ENTRY|[15]|System.debug(ApexLog): My debug message
Response	The row you provided from a Salesforce debug log indicates the following:
	- `26.0.1 (123456795)`: This part of the entry includes a timestamp and, potentially, information about the duration of the execution. The specific numbers and format may vary depending on your Salesforce instance and log configuration.
	- `SYSTEM_METHOD_ENTRY`: This indicates that a system method is being invoked. In this case, it's a system method related to logging.
	- `[15]`: This is typically a reference to the line number in your code where the system method is being called. In this case, it suggests that this entry corresponds to line 15 of your code.
	- `System.debug(ApexLog): My debug message`: This part of the entry specifies the details of the system method call. It shows that the `System.debug` method is being called with the argument `ApexLog`, which is a reference to an Apex log category, and the message "My debug message" that you are logging.
	Interpreting this log entry:
	- The entry indicates that your Apex code includes a call to the `System.debug` method to log a debug message. Debug messages are often used for troubleshooting and understanding the flow of your code.
	- The message "My debug message" is the custom message you are logging for debugging purposes.
	- This log entry is not an error or issue. It's a standard way of using debug statements to log information for diagnostic purposes while running your code.
	- You don't need to take any specific action about this entry unless you are actively troubleshooting or debugging your code and want to review the logged messages to identify issues or verify the behavior of your code.
	Overall, this log entry is part of the standard debugging process in Salesforce, and it helps developers track the execution and behavior of their code during runtime.

Please note: Debug messages in a production Salesforce org should generally be avoided or used sparingly. It's not considered a best practice to have debug messages enabled and visible in a production environment for several reasons:

- **Security and data privacy**: Debug messages may contain sensitive information about your organization's data, system configuration, or code logic. Making this information readily available in a production environment can pose security risks if unauthorized users gain access to it. It's essential to protect sensitive data and maintain data privacy compliance.

- **Performance impact**: Excessive debug messages can impact the performance of your production org, as they generate additional logging and consume system resources. This can slow down the execution of your code and affect the user experience.

- **Clutter and confusion**: Debug messages can clutter the logs and make it challenging to identify critical issues or errors when they occur. In a production environment, it's essential to maintain clean and meaningful logs for effective monitoring and troubleshooting.

Best practices for handling debug messages in a production Salesforce org include:

- **Conditional logging**: Use conditional statements to control when debug messages are generated and logged. You can create a system variable or custom setting that allows you to enable or disable debug mode. When debugging is not needed, turn off debug messages by setting the variable to `false`.

- **Log levels**: Salesforce provides different log levels (for example, DEBUG, INFO, ERROR) that allow you to control the verbosity of your logs. In production, use a higher log level (for example, INFO) to log only critical information and errors while suppressing less important debug messages.

- **Log to secure locations**: If you do need to log debug messages for troubleshooting in a production org, consider logging them to a secure and controlled location, such as a custom object or a designated logging platform, rather than displaying them in the **user interface** (**UI**).

In summary, debug messages should be used judiciously in a production Salesforce org to balance the need for troubleshooting and the importance of security, performance, and clean logs. Following best practices for conditional logging and log levels helps maintain a secure and efficient production environment.

The power of using ChatGPT to analyze Salesforce debug logs lies in its ability to swiftly decipher complex log entries, pinpoint issues, and offer actionable insights for resolution. ChatGPT can act as a knowledgeable partner in the debugging process, helping developers and administrators navigate the intricacies of log data. It can provide explanations, suggest potential causes of errors, recommend debugging strategies, and even offer code snippets for remediation. By leveraging ChatGPT's expertise, users can streamline their debugging efforts, accelerate issue resolution, and maintain the health and performance of their Salesforce orgs with confidence.

Real-time resolution using ChatGPT

An example of a Salesforce error that may appear to a user via the UI is the "Insufficient Privileges" error. This error message is displayed to users when they attempt to access or perform an action on a record or functionality for which they do not have the necessary permissions.

The error message typically reads something like this:

"Insufficient Privileges. You do not have the level of access necessary to perform the operation you requested. Please contact the administrator of this record for help."

This error occurs when a user tries to view, edit, delete, or perform any operation that requires specific access rights, such as read or edit access to a record, but their profile or permission settings do not grant them the required permissions. It's a security feature in Salesforce to prevent unauthorized access to sensitive data and functionalities.

In order to address this error a user with sufficient privileges, typically a system administrator, needs to update the permissions of the user getting the error. Let's see how ChatGPT can help:

Prompt	Please provide step-by-step directions for resolving the "Insufficient Privileges. You do not have the level of access necessary to perform the operation you requested. Please contact the administrator of this record for help." error for user JKubon.
Response	Resolving the "Insufficient Privileges" error for a specific user such as JKubon in Salesforce typically involves adjusting the user's profile or permission settings to grant the necessary access. Here are step-by-step directions to resolve this error for user JKubon: 1. Log in to Salesforce: Log in to your Salesforce administrator account or an account with administrative privileges. 2. Navigate to Setup: Click on your user profile icon at the top-right corner of the screen and select "Setup" from the drop-down menu. 3. Access User Management: In the left-hand sidebar, under "Administer", expand the "Users" section, and click on "Users". 4. Locate User JKubon: In the "Users" page, use the search or filter options to locate the user JKubon whose privileges need to be adjusted. 5. Edit User Profile: Click on JKubon's username to access their user details page. 6. Modify Profile or Permission Set: 7. Option 1: If JKubon's profile needs to be adjusted, scroll down to the "Profile" section and click on the "Edit Profile" button. Update the profile settings to grant the necessary access rights, such as read, edit, or delete permissions.

```
Option 2: If you are using permission sets to manage user
privileges, scroll down to the "Permission Sets" section
and click on "Manage Assignments." Assign the appropriate
permission sets to JKubon that grant the required access.

Save Changes: After making the necessary adjustments to
the profile or permission set, click the "Save" button to
save the changes.

Test Access: Log out of your administrator account and log
in as JKubon (or have JKubon log in). Try to access the
record or perform the operation that previously resulted
in the "Insufficient Privileges" error to ensure the issue
has been resolved.

Verify Access: Verify that JKubon now has the appropriate
access to the records and functionalities required for
their role.

12.  Provide Training (If Necessary): If JKubon's access
was restricted due to a lack of training or understanding,
consider providing training or guidance on how to use
Salesforce effectively.

By following these steps, you should be able to resolve
the "Insufficient Privileges" error for user JKubon and
ensure they have the necessary access rights within
Salesforce to perform their tasks without encountering
this error.
```

In many organizations, especially those with well-established Salesforce governance and compliance procedures, making changes to a user's permissions in a production Salesforce org often requires careful consideration, approvals, and adherence to change management processes. These processes are put in place to maintain the integrity and stability of the Salesforce environment, as hasty or unverified changes can have unintended consequences. Changes to user permissions may impact data security, compliance, and overall system functionality.

Before modifying user permissions, larger organizations typically follow a structured change management process that involves documenting the requested changes, obtaining appropriate approvals from stakeholders or a change advisory board, and scheduling the changes for deployment during a designated release window or maintenance window. This ensures that changes are thoroughly reviewed, tested, and executed in a controlled manner to minimize disruptions and mitigate risks.

In such environments, it's essential to collaborate closely with Salesforce administrators, compliance teams, and other relevant stakeholders to initiate and track changes to user permissions properly. By adhering to established change management protocols, organizations can maintain a stable and secure Salesforce environment while addressing user access needs in a structured and compliant manner.

Debugging best practices and continuous improvement

Leveraging ChatGPT in the context of debugging, troubleshooting, and adhering to best practices in Salesforce development offers substantial value to organizations. ChatGPT serves as a versatile and readily available resource that can help developers and administrators stay informed about the latest best practices, patterns, and anti-patterns in debugging and troubleshooting. Here's how ChatGPT can contribute to creating a top-notch Salesforce org:

1. **Access to best practices**: ChatGPT can provide instant access to the latest best practices in Salesforce debugging and troubleshooting. It can guide developers in identifying common pitfalls and suggest strategies to avoid them. This ensures that the Salesforce org is built with a strong foundation, reducing the likelihood of errors and performance issues down the road.

2. **Real-time guidance**: When developers encounter issues or errors, ChatGPT can assist in real time by offering suggestions for resolution. Whether it's interpreting error messages, recommending code changes, or guiding through the debugging process, ChatGPT can help streamline troubleshooting efforts and lead to quicker issue resolution.

3. **Debug log analysis**: ChatGPT's ability to interpret debug logs and explain the meaning of various entries provides a significant advantage. It can assist in identifying bottlenecks, governor limit breaches, and other performance-related issues, enabling developers to fine-tune their code for optimal performance.

4. **Encouraging best practices adherence**: By integrating ChatGPT into the development workflow, organizations can foster a culture of best practices adherence. Developers can rely on ChatGPT to remind them of established patterns and anti-patterns, ensuring that the code base remains clean, efficient, and scalable.

5. **Enhanced user experience**: A well-maintained Salesforce org leads to a better user experience. When users encounter minimal errors and experience fast and responsive performance, they are more likely to engage with the system enthusiastically and may even become advocates, sharing their positive experiences with peers.

ChatGPT's role in Salesforce debugging and troubleshooting goes beyond providing solutions to immediate problems. It empowers developers to proactively build high-quality solutions, stay informed about best practices, and maintain an org that not only meets user needs but also exceeds expectations. By leveraging ChatGPT's knowledge and guidance, organizations can create a Salesforce environment that users truly appreciate and recommend to others, ultimately contributing to their success and growth.

Summary

This chapter covered how ChatGPT has revolutionized the Salesforce debugging process. It focused on key aspects, including issue identification with ChatGPT's guidance, the intricacies of tracing and analyzing debug logs, and the capability of real-time resolution using ChatGPT. Moreover, the chapter emphasized the importance of adhering to debugging best practices and continuously improving debugging skills with ChatGPT as a valuable resource.

The next and concluding chapter of this book will embark on a journey of reflection and imagination, exploring the transformative impact of ChatGPT in the Salesforce development processes. The chapter will recap the various ways ChatGPT can enhance Salesforce project processes, covering aspects such as user-story creation, project documentation, debugging, and testing. The chapter goes on to delve into informed predictions about the future evolution of ChatGPT in the context of Salesforce, encouraging contemplation about how these advancements might revolutionize Salesforce development, administration, and the overall user experience.

14

What You've
Learned and What's Next

In this concluding chapter, we aim to summarize and reflect on the key insights you've acquired throughout this book. This chapter is organized into three main sections that cover various aspects of working with generative AI, such as ChatGPT, in the context of Salesforce and IT development.

The first part of this chapter reviews how to effectively engage with ChatGPT as a tool within the Salesforce environment. This includes treating ChatGPT as a knowledgeable partner in the development process, understanding its capabilities, and applying it across all stages of project activities. Whether it's assisting with documentation, design, grooming, development, or testing, we'll outline the practical ways in which ChatGPT can be utilized, and provide guidance on how to leverage it when challenges arise.

The middle section takes a closer look at the current state of generative AI and its impact on IT development. This includes an examination of the latest technological developments, current best practices, and existing challenges. The focus here is on the practical applications of AI in IT, the trends shaping the industry, and the lessons that can be drawn from contemporary use cases.

In the final part of this chapter, we'll consider the future of IT development concerning generative AI. This will involve us exploring potential developments, considering the ethical implications, and identifying areas for growth and innovation. The emphasis will be on informed predictions and thoughtful analysis, rather than speculative claims.

Overall, this chapter serves as both a recap of the learning journey and an exploration of the paths that lie ahead. It intends to provide a clear, grounded understanding of where we are now, and what the future may hold for integrating generative AI into the complex and ever-evolving landscape of IT development.

A journey recapped

Across this book, a handful of points have been made again and again:

- **Use ChatGPT across the project life cycle**: ChatGPT's application spans various project activities, including documentation, design, grooming, development, and testing within Salesforce projects. Developers should utilize it for drafting technical documents, modeling designs, aiding in requirement analysis, code development, and testing automation. This multifaceted engagement enables efficient project execution and quality enhancement.

- **Explore design alternatives with ChatGPT**: ChatGPT should be used to explore various design alternatives for Salesforce projects, including the evaluation of different options, their implications, and alignment with business needs. By engaging in detailed dialogues and posing complex queries, developers can leverage ChatGPT's capabilities to analyze and compare different solutions, ensuring a comprehensive consideration of the project requirements.

- **Utilize ChatGPT as a partner**: Salesforce developers should treat ChatGPT as a partner with extensive knowledge of the Salesforce ecosystem. During the development process, ChatGPT should be consulted for code reviews, suggestions on scalability, and troubleshooting complex challenges. This collaborative model fosters a dynamic environment for innovation and problem-solving.

- **Seek guidance from ChatGPT when you're stuck**: When faced with challenges or roadblocks, Salesforce developers should turn to ChatGPT for assistance in crafting prompts or generating solutions. By articulating the problem and seeking guided exploration, developers should be able to tap into ChatGPT's analytical capabilities to break down complex issues and derive actionable insights.

- **Avoid making assumptions about ChatGPT's limitations**: Developers should approach ChatGPT with an open mind and not make assumptions about what it cannot do. By engaging in iterative and exploratory dialogues, developers will discover new ways to leverage ChatGPT's capabilities in the Salesforce context, even in complex scenarios. This open exploration promotes a more nuanced understanding and utilization of generative AI in Salesforce development.

- **Balance ChatGPT's assistance with human oversight**: While ChatGPT acts as a valuable partner and accelerator, its collaboration with human experts, especially senior developers or architects, remains vital. Developers must recognize that ChatGPT can greatly aid the development process, but human judgment, creativity, and critical thinking are still needed, particularly in intricate design decisions that may require user stories and requirements to be revisited.

By focusing on these aspects, Salesforce developers can enhance their collaboration with ChatGPT, leveraging it as a powerful tool that complements human intelligence across various stages of project development. The synthesis of AI-driven insights with human creativity and judgment creates a process that fosters innovation, efficiency, and growth within the Salesforce landscape.

The current landscape of generative AI

The current state of generative AI in software development represents a significant evolution in automating and enhancing various aspects of the development life cycle. One of the key areas where generative AI has made a substantial impact is code generation and automation. These models can generate code snippets, templates, and even entire classes or functions by understanding the contextual requirements provided by developers. They can create code in multiple languages, aligning with coding standards and best practices, thus speeding up the development process and reducing common errors.

Furthermore, developers are engaging with AI models as virtual pair programmers, assisting in code reviews, proposing alternative solutions, and aiding in debugging. This integration of AI into the development process fosters a dynamic and responsive environment that encourages exploration and innovation. The automation of testing scenarios and the generation of test cases is another exciting application. AI can analyze the code base and requirements to create extensive test suites, ensuring early identification of bugs and freeing human resources for more complex tasks.

Generative AI is also becoming a valuable tool for personalized learning and skill development, providing tailored examples and guidance that adapt to individual needs and competencies. This facet of AI has implications for continuous learning, whether you're mastering new programming languages or complex algorithms. Another notable trend is the integration of AI into existing development tools and platforms, from IDE plugins to CI/CD pipelines, offering contextual assistance right where developers need it.

The advancement of generative AI in software development is not without its challenges. Ethical considerations and human oversight are vital to ensure compliance with legal, security, and ethical standards. Striking the right balance between automation and human creativity remains a nuanced challenge. The current state of generative AI in software development is marked by transformative changes that are reshaping traditional development paradigms. The collaboration between human intelligence and AI, which is guided by ethical considerations, remains central to unlocking the full potential of this technology, promising efficiency, quality improvements, creativity, exploration, and personalized growth for developers.

Conjectures on generative AI's evolution

In the coming years, generative AI is expected to significantly impact the software development landscape, transforming the way developers work, collaborate, and innovate. Enhanced productivity will be within reach as AI takes on routine coding tasks, freeing developers to focus on complex problems and creative solutions.

Learning and professional growth are poised to become more adaptive and personalized. Generative AI will offer guidance tailored to individual needs and industry trends, bridging the gaps between novice and expert, theory and practice.

Intelligent and adaptive development environments will emerge as powerful allies for developers. These environments will understand the coding context, predict potential challenges, and offer solutions proactively, facilitating a more efficient development process.

The domain of automated testing and quality assurance will be revolutionized by AI's ability to craft precise test cases and guard the quality of code. The relentless pursuit of perfection in code quality will be enabled by intelligent automation, with AI acting as both the creator and reviewer.

The ethical integration of AI will be a critical challenge, demanding responsible practices that ensure transparency, accountability, and trust. As AI takes on a more prominent role, ethical considerations will become paramount, shaping the frameworks within which AI operates.

Collaboration will transcend borders and barriers as AI enables a global community of developers to work together seamlessly. Open source innovation and real-time collaboration will become defining features of the new development ecosystem, enriched by diverse perspectives and collective wisdom.

Security in software development will be bolstered by AI's precise and vigilant mechanisms. The proactive identification of vulnerabilities and implementation of robust defenses will reinforce the trustworthiness of digital systems.

Strategic alignment with business goals will become more dynamic and responsive as AI supports real-time analysis and predictive modeling. This alignment will enable organizations to not only react to market changes but to anticipate and shape the future, aligning technology with strategic objectives.

The integration of generative AI into software development over the next 5 years is poised to reshape the field. From enhancing productivity to fostering collaboration and ensuring quality, the impact of AI will be profound but must be navigated with care and consideration regarding ethical, human, and business factors. The future promises a more efficient, responsive, and creative development landscape, guided by the intelligent and responsible use of AI technology.

The future of Salesforce development with generative AI

"In the vibrant cityscape of 2029, Sarah begins her day like many Salesforce developers of her generation, with a friendly greeting to her personal AI companion, AIDEN.

"Good morning, AIDEN. Let's plan our day," Sarah says, her voice filled with the anticipation of another exciting workday.

"Of course, Sarah! Your work AI assistant, CORA, just messaged me about a priority task to make an update to the CRM system. Would you like to start with that?" AIDEN responds, their voice cheerful and helpful.

With a nod, Sarah heads to her workspace, a comfortable blend of ergonomic design and cutting-edge technology. The commute to work is a thing of the past; her home office is seamlessly connected to all the corporate systems through AI interfaces.

Once settled, Sarah activates CORA, her work AI assistant, who immediately coordinates with AIDEN to ensure a smooth transition. "CORA, let's review that task. And please bring in FREDDY, the AI from Finance, and MIRA from Supply Chain. We'll need their insights."

As the developer's day unfolds, the lines between human and AI interaction blur. Each system's AI embodies specific roles and characteristics, from the assertive FREDDY in Finance to the meticulous MIRA in Supply Chain. They aren't mere tools but partners, each possessing their own personality and expertise.

The morning is filled with complex problem-solving, all conducted through speech and natural language. Reading and writing code has become a specialized skill, a relic of the past that only the old-timers cling to. In this world, communication is key, and the AIs assist, challenge, and innovate alongside the human team.

Sarah's lunch break is a collaboration with AIDEN, planning a weekend getaway, while CORA coordinates with the other AIs to ensure that work progress continues uninterrupted.

In the afternoon, a creative session ensues. Sarah, CORA, and the other AIs brainstorm solutions to enhance the CRM system's user experience. Their thoughts flow in a symphony of voices, a dynamic interplay of human intuition and AI-powered insights.

The day isn't without its challenges. An unexpected issue in the HR system requires immediate attention, and LUCIA, the HR AI, joins the team. Together, they diagnose the problem, develop a solution, and even improve the system's efficiency.

A highlight of the day is the mentoring session with a junior developer, James. Sarah and her AI companions guide him through a complex task, the AIs acting as facilitators, providing resources, examples, and encouragement.

As the workday concludes, Sarah reflects on the profound change in her role. Coding, once the core of her profession, has evolved into orchestrating a collaborative dance between human and AI intellects. The AIs are not just pieces of software but colleagues, each contributing unique value.

She bids farewell to CORA and the other work AIs and spends her evening with AIDEN, exploring hobbies, learning new recipes, and simply enjoying a conversation about life and technology.

The future of Salesforce development, as envisioned in Sarah's world, is an inspiring blend of humanity and technology. The AIs are more than tools; they are partners and friends, each with a unique presence. The tasks are not confined to screens and keyboards but transcend into an organic collaboration of voices, ideas, and creativity.

In 5 years, the profession of "Salesforce Developer" will transform into something richer, more accessible, and filled with endless possibilities. The developer's role will shift from configuration and coding to guiding intelligent, anthropomorphized systems, unlocking potentials that we can only begin to imagine today. The essence of development will become an art of communication, innovation, and empathy, a future where technology and humanity walk hand in hand.

Appendix A:
Case Study

We use a case study to tie the topics in this book together. It is a real-world example of a Salesforce project that requires the techniques and skills taught in this book.

Problem statement

Acme Corp has recently deployed Salesforce Sales Cloud, and as part of that project, they created an integration with Oracle ERP. This integration keeps the Salesforce and Oracle ERP accounts in sync. Acme also has Snowflake in production. Snowflake is consuming data from multiple internal and external systems and has started to ramp up AI-based analytics. The Acme business leadership has asked the IT department to design a solution and estimate the time and cost of what it would take to have a **Snowflake** section on the Salesforce account pages that summarizes the information in Snowflake for the account, and also allows the sales rep to select a contact related to the account and automatically copy the Snowflake information to the email.

Assumptions

Let's look at the assumptions for this case study.

Current state

- There is an existing Sales Cloud implementation that's currently used by sales reps

- The Account and Contact objects are being used by sales reps

- There is an existing integration between Salesforce and Oracle ERP

- No middleware is in use, so integrations are point-to-point

- No integration exists today between Salesforce and Snowflake

Future state

- There is valuable data available within their Snowflake platform that is not visible to the sales reps today. The request is to update the account page to display the following information:

  - Billing and payment history analysis (Snowflake will provide a one-paragraph natural language summary)

  - Products quoted but not purchased (Snowflake will provide a one-paragraph natural language summary)

  - Customer sentiment analysis data (Snowflake will provide a three to five-paragraph natural language summary)

  - Industry and market trends that are impacting the account (Snowflake will provide a three to five-paragraph natural language summary)

  - Recent news and events related to the customer (Snowflake will provide a three to five-paragraph natural language summary)

- The sales rep can click on a button saying that the information from Snowflake is incorrect, provide a written explanation, and click a button to send the explanation back to Snowflake.

- The sales rep can click on a button to post a summary of the account and Snowflake information to Slack, add commentary, and then post it to Slack.

- The sales rep can run a report with a summary of the account and Snowflake information.

With the problem statement and assumptions defined, let's start by working with ChatGPT to define and write the epics.

Creating epics

The first step is to have ChatGPT digest the problem statement and assumptions:

Prompt	I am going to ask you a series of questions so that you can write epics and user stories for a project to integrate Salesforce and Snowflake. The first step is for you to ingest the problem statement and assumptions. Please do not write a response attempting to "answer" my prompt; simply ingest the materials and prepare to be asked questions. \<The problem statement and assumptions mentioned previously\>
Response	Certainly! I will await your questions related to the problem statement and assumptions. Thank you for providing the context and setting the direction for our collaboration. Feel free to proceed with your inquiries when you are ready.

The next step is to have ChatGPT make a first pass at writing the epics:

Prompt	Please suggest the epics that capture the intent of this problem statement and assumptions.
Response	Certainly! The situation presented by Acme Corp requires a thoughtful and comprehensive approach. Based on the problem statement and assumptions, the following epics capture the intent of the proposed enhancement to the Salesforce account pages, aligning with both the existing architecture and the new requirements:

1. Integrate Snowflake data with Salesforce account pages:

   - Design and implement a point-to-point integration between Salesforce and Snowflake
   - Ensure real-time or near-real-time synchronization of account-specific data

2. Display Snowflake information on Salesforce account pages:

   - Create a dedicated "Snowflake" section within the Salesforce account pages
   - Provide a visualization of various data points, such as billing history, product analysis, customer sentiment, industry trends, and recent news/events

3. Email functionality with Snowflake data:

   - Enable sales reps to select a contact and automatically attach the Snowflake information summary to an email

4. Feedback mechanism for incorrect Snowflake information:

   - Implement a button and an input field allowing sales reps to mark information as incorrect and send feedback to Snowflake

5. Integrate the Snowflake summary with Slack:

   - Design a feature enabling sales reps to post a summary of the account and Snowflake information to Slack, including the ability to add commentary

6. Provide reporting capabilities for the account and Snowflake information:

   - Develop reporting functionality within Salesforce to generate summaries of the account and Snowflake information as required by sales reps

7. Compliance, security, and performance considerations:

   - Assess and address the security, performance, and compliance requirements pertinent to integrating and displaying Snowflake information within Salesforce

<table>
<tr><td></td><td>

8. User training and documentation:

- Develop comprehensive training materials and documentation to ensure that sales reps can effectively utilize the new features

These epics encapsulate a robust plan that meets the business objectives while also considering important technical factors such as security and performance. By following this framework, the IT department at Acme can work in collaboration with sales reps to innovate and enhance the existing Salesforce implementation, thus fulfilling the strategic vision articulated by the business leadership.

</td></tr>
</table>

Let's analyze each of the proposed epics, review how well they match the problem statement and assumptions, and craft a prompt to refine the initial response.

1. Integrate Snowflake data with Salesforce account pages:

- Design and implement a point-to-point integration between Salesforce and Snowflake
- Ensure real-time or near-real-time synchronization of account-specific data

The first bullet is good, but more detail is needed. Specifically, it should say what the integration will do – that is, it will be used for Salesforce to get information about the account from Snowflake. The second bullet is not correct. The call will be made when a Salesforce user views the account page. Because some of the information provided by Snowflake can change fairly quickly, each time the page is viewed, another call will be made to Snowflake. For a Salesforce org with many users who view the same account page multiple times a day, the epic may have the Snowflake calls manually initiated by the users. For this example, however, each account page refresh will refresh the Snowflake information.

2. Display Snowflake information on Salesforce account pages:

- Create a dedicated "Snowflake" section within the Salesforce account pages
- Provide visualization of various data points, such as billing history, product analysis, customer sentiment, industry trends, and recent news/events

This is good. The user stories will include the details of the information to be displayed and how to display the information.

3. Email functionality with Snowflake data:

- Enable sales reps to select a contact and automatically attach the Snowflake information summary to an email

This is a good starting point, but the language should be updated to have the Snowflake information in the body of the email.

4. Feedback mechanism for incorrect Snowflake information:

- Implement a button and an input field allowing sales reps to mark information as incorrect and send feedback to Snowflake

This is good.

5. Integrate the Snowflake summary with Slack:

- Design a feature that enables sales reps to post a summary of the account and Snowflake information to Slack, including the ability to add commentary

This is good. One quirk to note about ChatGPT is that it uses the writing rule about using the same words to say the same thing over and over. As we can see, the bullets start with *Enable, Implement*, and *Design*. Technical writing should not follow this rule, but the authors have yet to figure out how to get ChatGPT to follow technical writing rules.

6. Provide reporting capabilities for the account and Snowflake information:

- Develop reporting functionality within Salesforce to generate summaries of the account and Snowflake information as required by sales reps

This is good.

7. Compliance, security, and performance considerations:

- Assess and address the security, performance, and compliance requirements pertinent to integrating and displaying Snowflake information within Salesforce

There are times when ChatGPT will extend/augment the provided directions. The problem statement didn't include **non-functional requirements (NFRs)** but ChatGPT, quite correctly, is suggesting an epic for the needed NFRs.

8. User training and documentation:

- Develop comprehensive training materials and documentation to ensure that sales reps can effectively utilize the new features

While the problem statement did not include the development of training and documentation materials, ChatGPT is correct in suggesting this as an epic.

The next step is to craft a prompt to refine ChatGPT's first pass at the epics. The prompt needs to request revisions to epics *1. Integrate Snowflake data with Salesforce account pages* and *3. Email functionality with Snowflake data*:

Prompt	Please rewrite the first bullet for epic 1 to note that the integration will be used for Salesforce to call Snowflake to get information for a specific Salesforce account when a user views the account page. Please remove the second bullet. Please rewrite the first bullet for epic 3 to have the Snowflake data in the body of the email. Other than these changes, do not change any of the other language in the first response.

Response	Certainly! Here are the revised epics with the specified changes:
	
	1. Integrate Snowflake data with Salesforce account pages:
	
	   • Design and implement a point-to-point integration for Salesforce to call Snowflake and retrieve information for a specific Salesforce account when a user views the account page
	
	2. Display Snowflake information on Salesforce account pages:
	
	   • Create a dedicated "Snowflake" section within the Salesforce account pages
	
	   • Provide visualization of various data points such as billing history, product analysis, customer sentiment, industry trends, and recent news/events
	
	3. Email functionality with Snowflake data:
	
	   • Enable sales reps to select a contact and automatically insert the Snowflake information summary into the body of an email
	
	   • Implement the email functionality aligned with the requirements
	
	4. Feedback mechanism for incorrect Snowflake information:
	
	   • Implement a button and an input field allowing sales reps to mark information as incorrect and send feedback to Snowflake
	
	5. Integrate the Snowflake summary with Slack:
	
	   • Design a feature enabling sales reps to post a summary of the account and Snowflake information to Slack, including the ability to add commentary
	
	6. Provide reporting capabilities for the account and Snowflake information:
	
	   • Develop reporting functionality within Salesforce to generate summaries of the account and Snowflake information as required by sales reps
	
	7. Compliance, security, and performance considerations:
	
	   • Assess and address the security, performance, and compliance requirements pertinent to integrating and displaying Snowflake information within Salesforce

	8. User training and documentation:
	• Develop comprehensive training materials and documentation to ensure that sales reps can effectively utilize the new features
	These modifications emphasize the specific interaction between Salesforce and Snowflake, focusing on the real-time retrieval of information for individual Salesforce accounts and the inclusion of Snowflake data in the body of the email.

Now that there is a good set of epics, the next step is to create user stories for them.

User stories

ChatGPT works best when used as a writing partner. Do not try to write a perfect prompt the first time. Leave some possibilities open and allow ChatGPT to make suggestions. **Large language models (LLMs)** are, at their core, prediction machines. Start the process of creating user stories from epics by allowing ChatGPT some freedom to predict what might work.

Before we start working on the user stories, let's define the user story structure that will be used:

Prompt	In a well-structured agile user story, the narrative typically begins with a clear statement of the goal, following the pattern "As a [user type], I want [an action] so that [benefit/value]." The simplicity of this format aids in aligning the development work with user needs and business goals. Alongside the story, the acceptance criteria define the specific conditions that the functionality must satisfy to be accepted by the stakeholder. Here, employing the Gherkin syntax elevates the precision and readability of the criteria. Written in a natural, business-readable language, the Gherkin syntax follows a "Given-When-Then" pattern: "Given [a context], When [an action is carried out], Then [a set of observable consequences should occur]." Please follow this process when writing user stories from epics.
	Please do not write a response attempting to "answer" my prompt; simply ingest the materials and prepare to be asked to write user stories.
Response	Understood. I have taken note of your guidelines for structuring user stories and employing the Gherkin syntax for defining acceptance criteria. I am prepared to assist with writing user stories as and when you require. Feel free to proceed with your next inquiry.

Let's briefly touch on the process of decomposing user stories from **Level 1** (**L1** – high level) to the level where they can be groomed, estimated, and scheduled for development.

Decomposing user stories

When writing user stories from epics, it is important to understand the process of decomposing user stories from L1 to a level where they become actionable:

- **Level 1 (L1)**: These are high-level user stories that often encapsulate the broad vision or strategic goals of a project. L1s are mainly oriented toward understanding the customer's major needs or the overall business objective.

- **Level 2 (L2)**: This level further decomposes L1s into more detailed functionalities or features. L2s describe significant functionalities that must be implemented to fulfill the strategic goals, yet may not be detailed enough for development.

- **Level 3 (L3)**: These stories break down L2s into finer details. L3s are closer to the actual development stage, with enough detail to understand the specific requirements, but may still need further refinement.

- **Level 4 (L4)**: L4s are the most granular level of user stories, ready for the development stage. They include all the necessary details and acceptance criteria, and they are small enough to be tackled within a single iteration or sprint.

The decomposition process takes a top-down approach. Starting from L1, each level is dissected into the next, focusing on more granular details. From L1 to L2, the broad vision is translated into key functionalities or feature sets. From L2 to L3, these functionalities are further detailed into specific requirements. The decomposition from L3 to L4 refines these requirements into actionable user stories, complete with acceptance criteria, estimates, and all the necessary details for coding and testing.

Determining when a user story has been decomposed enough for grooming for development is a crucial part of the process. A user story is considered ready when it is concise and clear, includes well-defined acceptance criteria, and is small enough to be implemented within a sprint. One way to determine this is to apply the **Independent**, **Negotiable**, **Valuable**, **Estimable**, **Small**, and **Testable** (**INVEST**) criteria.

With ChatGPT prepared to write user stories, let's get started.

Integrate Snowflake data with Salesforce account pages

Prompt ChatGPT to write the user stories for the first epic:

Prompt	Please write the user stories for this epic:
	1. Integrate Snowflake data with Salesforce account pages:
	• Design and implement a point-to-point integration for Salesforce to call Snowflake and retrieve information for a specific Salesforce account when a user views the account page

Response	Certainly! Here are the user stories for the specified epic, structured as per your guidance:
	User story 1 – retrieve Snowflake data for a specific Salesforce account:
	• Goal: As a sales rep, I want Salesforce to call Snowflake and retrieve information for a specific Salesforce account when I view the account page so that I have real-time insights related to that account
	• Acceptance criteria:
	▪ Given a sales rep is viewing a specific Salesforce account page
	▪ When the page is loaded
	▪ Then the system should call Snowflake and retrieve the relevant information for that specific account
	This user story aligns with the goals of the given epic, highlighting both the functionality of the integration and the timeliness of the information retrieval, and it provides a clear and actionable framework for development.

Please note that ChatGPT does not replace Salesforce developers! ChatGPT is an assistant and an accelerator. Using ChatGPT for Salesforce projects still requires understanding the work to be done and being able to provide guidance.

In the case of user story 1 – *retrieve Snowflake data for a specific Salesforce account*, it is important to understand how to decompose an L1 story to the L2 or L3 level needed to apply the INVEST criteria. In the case of user story 1, for example, an Apex Call Out will be needed:

Prompt	Please decompose this user story to L2 stories that meet the INVEST criteria for Salesforce development. The L2 user stories should cover creating a secure Snowflake connection and the Call Out to request information from Snowflake. There should also be a user story to address error handling.
	User story 1 – retrieve Snowflake data for a specific Salesforce account:
	• Goal: As a sales rep, I want Salesforce to call Snowflake and retrieve information for a specific Salesforce account when I view the account page so that I have real-time insights related to that account
	• Acceptance criteria:
	▪ Given a sales rep is viewing a specific Salesforce account page
	▪ When the page is loaded
	▪ Then the system should call Snowflake and retrieve the relevant information for that specific account

Response	Certainly! Based on your requirements, here are the L2 user stories decomposed from the original user story, focusing on creating a secure Snowflake connection, the Call Out to request information, and error handling:
	
	L2 user story 1.1 – establish a secure connection with Snowflake:
	
	• Goal: As a sales rep, I want Salesforce to establish a secure connection with Snowflake when I view an account page, ensuring data integrity and privacy
	
	• Acceptance criteria:
	
	    ▪ Given a sales rep is viewing a specific Salesforce account page
	
	    ▪ When the page is loaded
	
	    ▪ Then the system should establish a secure connection with Snowflake
	
	• Notes: Specific, focused on security, and meets the INVEST criteria
	
	L2 user story 1.2 – call out to request specific account information from Snowflake:
	
	• Goal: As a sales rep, I want Salesforce to make a Call Out to Snowflake to request specific information related to a Salesforce account when I view the account page, providing me with timely insights
	
	• Acceptance criteria:
	
	    ▪ Given a sales rep is viewing a specific Salesforce account page and a secure connection to Snowflake has been established
	
	    ▪ When the page is loaded
	
	    ▪ Then the system should make a Call Out to Snowflake to retrieve the relevant information for that specific account
	
	• Notes: Specific, focused on the Call Out mechanism, and meets the INVEST criteria
	
	L2 user story 1.3 – implement error handling for Snowflake connection and data retrieval:
	
	• Goal: As a sales rep, I want any errors that occur during the connection to Snowflake or data retrieval to be handled gracefully so that I am informed about the issue and can continue using the system without disruption
	
	• Acceptance criteria:
	
	    ▪ Given a sales rep is attempting to view a specific Salesforce account page,
	
	    ▪ When an error occurs in connecting to Snowflake or retrieving data

	• Then the system should log the error, notify the sales rep, and allow them to continue using the system without disruption
	• Notes: Specific, focused on error handling, and meets the INVEST criteria
	These L2 user stories break down the original user story into specific, actionable tasks that adhere to the INVEST criteria. They provide a clear path for development, aligning with the best practices for Salesforce development, and ensuring that all essential aspects of the integration with Snowflake, including security and error handling, are addressed.

As ChatGPT noted, these L2 user stories meet the INVEST criteria and are therefore ready to be groomed and estimated.

Display Snowflake information on Salesforce account pages

Next is the user story for displaying the Snowflake information. The prompt will use the techniques we reviewed when working with the first epic:

| Prompt | Please write the user stories for the following epic. For each user story that does not meet the INVEST criteria, iteratively decompose the user stories until they do meet the INVEST criteria. These stories should cover writing a custom Lightning Web Component that has a separate text field for each of the categories of information from Snowflake, such as billing history, product analysis, and so on.

2. Display Snowflake information on Salesforce account pages:

• Create a dedicated "Snowflake" section within the Salesforce account pages

• Provide visualization of various data points, such as billing history, product analysis, customer sentiment, industry trends, and recent news/events |
|---|---|
| Response | Certainly! The epic you provided focuses on creating a custom Lightning Web Component within Salesforce to display various Snowflake data points on account pages. Here are the user stories, decomposed to meet the INVEST criteria.

User story 1 – create a Snowflake section within the Salesforce account pages:

• Goal: As a sales rep, I want a dedicated "Snowflake" section within Salesforce account pages so that I can quickly access the Snowflake data related to an account. |

- Acceptance criteria:

  · Given a sales rep is viewing a Salesforce account page

  · When the page is loaded

  · Then a dedicated "Snowflake" section should be displayed

L2 user story 1.1 – design the Snowflake section's layout:

- Goal: As a sales rep, I want the "Snowflake" section to have a consistent and user-friendly layout, allowing for clear visibility of different data categories

- Acceptance criteria:

  · Given a sales rep is viewing the "Snowflake" section

  · When the page is loaded

  · Then the layout of the section should be visually appealing and intuitive

User story 2 – display billing history in the Snowflake section:

- Goal: As a sales rep, I want to see the billing and payment history analysis from Snowflake within the "Snowflake" section so that I can understand the account's financial status

- Acceptance criteria:

  · Given a sales rep is viewing the "Snowflake" section

  · When the billing history data is retrieved from Snowflake

  · Then the billing history should be displayed in a dedicated text field

User story 3 – display product analysis in the Snowflake section:

- Goal: As a sales rep, I want to see products quoted but not purchased, as analyzed by Snowflake within the "Snowflake" section, so that I can understand opportunities for up-selling or cross-selling

- Acceptance criteria:

  · Given a sales rep is viewing the "Snowflake" section

  · When the product analysis data is retrieved from Snowflake

  · Then the analysis should be displayed in a dedicated text field

> User stories 4 to 6 – display customer sentiment, industry trends, and recent news/events
>
> Similarly, user stories for displaying customer sentiment analysis, industry trends impacting the account, and recent news/events related to the customer would be written. Each user story will define the specific requirements for displaying these data points within individual text fields.
>
> These user stories align with the INVEST criteria, ensuring they are specific, actionable, and relevant to Salesforce development. They guide the creation of a custom Lightning Web Component to provide a dedicated view of Snowflake data on Salesforce account pages, enhancing the experience for sales reps.

While it is tempting to anthropomorphize ChatGPT when reading user stories 4 to 6 in the preceding response – for example, thinking that ChatGPT is saying "Seriously, dude, do I need to write the same user story over and over?" it is important to recall that ChatGPT has been trained mostly on content from the internet, and this is what many people would do when faced with writing essentially the same passage over and over. ChatGPT is just a reflection of human behavior.

At the same time, separate user stories are needed here:

Prompt	Please write complete freestanding user stories for user stories 4 to 6.
Response	Certainly! Here are the complete freestanding user stories for items 4 to 6, focusing on the display of customer sentiment, industry trends, and recent news/events within the Snowflake section on the Salesforce account pages.
	User story 4 – display customer sentiment analysis in the Snowflake section:
	• Goal: As a sales rep, I want to see the customer sentiment analysis data from Snowflake within the "Snowflake" section so that I can gauge customer satisfaction and adapt my approach accordingly
	• Acceptance criteria:
	▪ Given a sales rep is viewing the "Snowflake" section
	▪ When the customer sentiment analysis data is retrieved from Snowflake
	▪ Then the sentiment analysis should be displayed in a dedicated text field

User story 5 – display industry trends in the Snowflake section:

- Goal: As a sales rep, I want to see the industry and market trends that are impacting the account, as analyzed by Snowflake, within the "Snowflake" section, so that I can strategize based on the macroeconomic environment

- Acceptance criteria:

  - Given a sales rep is viewing the "Snowflake" section

  - When the industry and market trend data is retrieved from Snowflake

  - Then the trends should be displayed in a dedicated text field

User story 6 – display recent news and events in the Snowflake section:

- Goal: As a sales rep, I want to see recent news and events related to the customer, as analyzed by Snowflake, within the "Snowflake" section, so that I can stay informed about significant developments

- Acceptance criteria:

  - Given a sales rep is viewing the "Snowflake" section

  - When the recent news and events data is retrieved from Snowflake

  - Then the news and events should be displayed in a dedicated text field

These user stories are tailored to guide the development of specific features within the custom Lightning Web Component on Salesforce account pages. By providing detailed goals and acceptance criteria, they align with the INVEST principles, ensuring that the development work is clear, feasible, and closely aligned with user needs and business objectives.

As is obvious from this prompt, each ChatGPT "chat" contains the context of the previous/response exchanges.

Email functionality with Snowflake data

Let's consider the prompt for this aspect:

| Prompt | Please write the user stories for the following epic. For each user story that does not meet the INVEST criteria, iteratively decompose the user stories until they do meet the INVEST criteria. These stories should cover adding an "Email" button to the Snowflake Lightning Web Component on the account page, developing a Flow-based wizard that prompts the user to select a contact to send the email to, and giving the user the ability to edit the email subject and body before sending it. |

Response	Certainly! Let's break down the epic into user stories that cover the requirements and meet the INVEST criteria.
	User story 1 – add an "Email" button to the Snowflake Lightning Web Component:
	• Goal: As a sales rep, I want an "Email" button within the Snowflake section of the account page so that I can initiate the process of sending Snowflake information via email
	• Acceptance criteria:
	• Given a sales rep is viewing the Snowflake section on the account page
	• When the sales rep looks at the Snowflake section
	• Then an "Email" button should be visible and clickable
	User story 2 – Create a Flow-Based wizard to select a contact:
	• Goal: As a sales rep, I want a wizard to guide me through the process of selecting a contact to send the email to so that I can choose the correct recipient
	• Acceptance criteria:
	• Given a sales rep clicks the "Email" button in the Snowflake section
	• When the wizard starts
	• Then the sales rep should be prompted to select a contact associated with the account
	User story 3 – allow editing of email subject and body:
	• Goal: As a sales rep, I want the ability to edit the email subject and body before sending the Snowflake information summary so that I can personalize and review the email content
	• Acceptance criteria:
	• Given a sales rep has selected a contact through the wizard,
	• When the email draft is presented
	• Then the sales rep should have the option to edit the email subject and body
	User story 4 – insert Snowflake information into the email's body:
	• Goal: As a sales rep, I want the Snowflake information summary to be automatically inserted into the body of the email so that I can quickly share insights without manual effort

- Acceptance criteria:

  - Given a sales rep is editing the email content

  - When the email draft is first presented

  - Then the Snowflake information summary for the specific account should be inserted into the body of the email

User story 5 – send the email to the selected contact:

- Goal: As a sales rep, I want to send the email with the Snowflake information to the selected contact so that the information reaches the intended recipient

- Acceptance criteria:

  - Given a sales rep has edited the email and is ready to send it

  - When the sales rep clicks the "Send" button

  - Then the email should be sent to the selected contact with the Snowflake information included

These user stories are meticulously crafted to cover the different aspects of the email functionality within the Snowflake section of the Salesforce Account page. By addressing each step and allowing for iterations, they guide the development work in alignment with user needs and follow the INVEST criteria to ensure independent, negotiable, valuable, estimable, small, and testable units of work.

Feedback mechanism for incorrect Snowflake information

Here's the prompt:

Prompt	Please write the user stories for the following epic. For each user story that does not meet the INVEST criteria, iteratively decompose the user stories until they do meet the INVEST criteria. These stories should cover adding a "Feedback" button to the Snowflake Lightning Web Component on the account page, developing a Flow-based wizard that allows the user to enter feedback about what Snowflake information is incorrect, calling an Apex Call Out to send the feedback to Snowflake, and the Apex Call Out.
	4. Feedback mechanism for incorrect Snowflake information:
	• Implement a button and an input field allowing sales reps to mark information as incorrect and send feedback to Snowflake

Response	Certainly! Here are the user stories for the epic "Feedback mechanism for incorrect Snowflake information," decomposed to meet the INVEST criteria.
	User story 1 – add a "Feedback" button to the Snowflake LWC:
	• Goal: As a sales rep, I want a "Feedback" button on the Snowflake Lightning Web Component on the account page so that I can report incorrect information
	• Acceptance criteria:
	• Given a sales rep is viewing the Snowflake information on the account page
	• When the sales rep wants to provide feedback
	• Then a "Feedback" button should be visible and clickable
	User story 2 – develop a Flow-based wizard for feedback:
	• Goal: As a sales rep, I want a Flow-based wizard to guide me through entering feedback about incorrect Snowflake information so that I can provide specific details easily
	• Acceptance criteria:
	• Given a sales rep has clicked the "Feedback" button
	• When entering feedback
	• Then a Flow-based wizard should guide the sales rep through the process of entering specific details about what information is incorrect
	User story 3 – create an Apex Call Out to send feedback to Snowflake:
	• Goal: As a sales rep, I want the feedback about incorrect Snowflake information to be sent to Snowflake so that the necessary corrections can be made
	• Acceptance criteria:
	• Given a sales rep has entered feedback through the Flow-based wizard
	• When the feedback is submitted
	• Then an Apex Call Out should send the feedback details to Snowflake for review
	User story 4 – handle errors in the Apex Call Out:
	• Goal: As a sales rep, I want to be informed if there's an error in sending my feedback to Snowflake so that I know whether my feedback was successfully sent or not

	• Acceptance criteria:
	
	    o Given a sales rep has submitted feedback through the Apex Call Out
	
	    o When there's an error in sending the feedback to Snowflake
	
	    o Then the sales rep should receive a clear error message indicating what went wrong
	
	User story 5 – confirmation of feedback sent:
	
	• Goal: As a sales rep, I want confirmation that my feedback about incorrect Snowflake information has been successfully sent so that I know it will be reviewed
	
	• Acceptance criteria:
	
	    o Given a sales rep has submitted feedback about incorrect Snowflake information
	
	    o When the feedback is successfully sent to Snowflake
	
	    o Then the sales rep should receive a confirmation message that the feedback has been sent and will be reviewed
	
	Through these decomposed user stories, the functionality of marking information as incorrect and sending feedback to Snowflake is thoroughly covered, meeting the INVEST criteria and aligning with real-world Salesforce development practices.

Integrate the Snowflake summary with Slack

Here's the prompt:

Prompt	Please write the user stories for the following epic. For each user story that does not meet the INVEST criteria, iteratively decompose the user stories until they do meet the INVEST criteria. These stories should cover adding a "Slack" button to the Snowflake Lightning Web Component on the account page, developing a Flow to call Slack, and posting a summary of the account information and the Snowflake information to Slack. 5. Integrate the Snowflake summary with Slack: • Design a feature enabling sales reps to post a summary of the account and Snowflake information to Slack, including the ability to add commentary

Response	Certainly! The following user stories break down the features and functionality required for the integration of the Snowflake summary with Slack, meeting the INVEST criteria.
	User story 1: Add a "Slack" button to the Snowflake LWC:
	• Goal: As a sales rep, I want a "Slack" button on the Snowflake Lightning Web Component on the account page so that I can share information with Slack
	• Acceptance criteria:
	• Given a sales rep is viewing the Snowflake information on the account page
	• When the sales rep wants to share this information with Slack
	• Then a "Slack" button should be visible and clickable
	User story 2 – develop a Flow-based input for Slack sharing:
	• Goal: As a sales rep, I want a Flow-based input process to allow me to compose a summary of the account and Snowflake information, along with my commentary, so that I can customize what I share with Slack
	• Acceptance criteria:
	• Given a sales rep has clicked the "Slack" button
	• When preparing to share information with Slack
	• Then a Flow-based input process should guide the sales rep through composing a summary and adding any commentary
	User story 3 – call the Slack API to post a summary and commentary:
	• Goal: As a sales rep, I want the composed summary and commentary of the account and Snowflake information to be posted to Slack so that my team can access it there
	• Acceptance criteria:
	• Given a sales rep has composed the summary and commentary through the Flow-based input process
	• When the sales rep submits the information
	• Then a call to the Slack API should post the summary and commentary to the specified Slack channel
	User story 4 – error handling for Slack integration:
	• Goal: As a sales rep, I want to be informed if there's an error in posting my summary and commentary to Slack so that I know whether the information was successfully shared or not

- Acceptance criteria:

  - Given a sales rep has submitted the summary and commentary to Slack

  - When there's an error in posting the information

  - Then the sales rep should receive a clear error message indicating what went wrong

User story 5 – confirmation of a successful Slack post:

- Goal: As a sales rep, I want confirmation that my summary and commentary of the account and Snowflake information have been successfully posted to Slack so that I know my team will see it

- Acceptance criteria:

  - Given a sales rep has submitted the summary and commentary to Slack

  - When the information is successfully posted

  - Then the sales rep should receive a confirmation message that the information has been shared with Slack

Through the iterative decomposition of user stories, this approach ensures that all aspects of the integration, from the user interface to the functional interaction with Slack's API, are addressed. It also considers error handling and user feedback, providing a comprehensive solution aligned with best practices in Salesforce development.

Providing reporting capabilities for the account and Snowflake information

Here's the prompt:

Prompt	Please write the user stories for the following epic. For each user story that does not meet the INVEST criteria, iteratively decompose the user stories until they do meet the INVEST criteria. These stories should include creating a parameterized report that is for one account and a button on the account page that invokes the report.
	6. Reporting capabilities for the account and Snowflake information:
	<ul><li>Develop reporting functionality within Salesforce to generate summaries of the account and Snowflake information as required by sales reps</li></ul>

Response	Certainly! The following user stories describe the functionality required for developing the reporting capabilities for the account and Snowflake information within Salesforce, with detailed adherence to the INVEST criteria.
	User story 1 – design a parameterized report template:
	• Goal: As a sales rep, I want a parameterized report template that can generate summaries for specific account and Snowflake information so that I can view the details for individual accounts
	• Acceptance criteria:
	· Given a sales rep wants to view a summary report for a specific account
	· When the sales rep selects the account
	· Then the system should provide a template that dynamically populates the summary based on the selected account's details
	User story 2 – create a "Generate Report" button on the account page:
	• Goal: As a sales rep, I want a "Generate Report" button on the account page so that I can easily initiate the creation of the summary report
	• Acceptance criteria:
	· Given a sales rep is viewing a specific account page
	· When wanting to create a summary report
	· Then a "Generate Report" button should be visible and clickable
	User story 3 – implement functionality to invoke the report:
	• Goal: As a sales rep, I want the "Generate Report" button to invoke the creation of the parameterized report for the specific account I'm viewing so that the report reflects the current account and Snowflake information
	• Acceptance criteria:
	· Given a sales rep has clicked the "Generate Report" button on the account page
	· When the button is clicked
	· Then the system should invoke the parameterized report, populating it with the specific account and Snowflake information

	User story 4 – enable report export options:
	• Goal: As a sales rep, I want the ability to export the generated report in various formats (for example, PDF and Excel) so that I can share or archive the information as needed
	• Acceptance criteria:
	▪ Given a sales rep has generated a report for a specific account
	▪ When wanting to export the report
	▪ Then options for exporting in various formats should be available
	User story 5 – error handling for report generation:
	• Goal: As a sales rep, I want to be notified if there's an error in generating or exporting the report so that I am aware of any issues that might occur
	• Acceptance criteria:
	▪ Given a sales rep has invoked the report generation
	▪ When there's an error in generating or exporting the report
	▪ Then the sales rep should receive a clear error message indicating what went wrong
	By breaking down the epic into these detailed user stories, we ensure that the reporting functionality is comprehensively addressed. This includes not only the core functionality of creating and invoking the report but also the user experience aspects, such as exporting and error handling, all of which align with Salesforce development practices.

Compliance, security, and performance considerations

Here's the prompt:

Prompt	Please write the user stories for the following epic. For each user story that does not meet the INVEST criteria, iteratively decompose the user stories until they do meet the INVEST criteria.
	7. Compliance, security, and performance considerations:
	• Assess and address the security, performance, and compliance requirements pertinent to integrating and displaying Snowflake information within Salesforce

| Response | Certainly! The epic focused on compliance, security, and performance considerations will necessitate a multifaceted approach. Here are the user stories that collectively capture the required functionalities.

User story 1 – conduct a security assessment:

- Goal: As a security analyst, I want to perform a comprehensive security assessment of the integration between Snowflake and Salesforce so that any vulnerabilities can be identified and addressed
- Acceptance criteria:
    - Given a security analyst is assessing the integration
    - When examining the integration
    - Then all potential security risks and vulnerabilities should be identified

User story 2 – implement security measures:

- Goal: As a developer, I want to implement the security measures identified from the assessment so that the integration is safeguarded against potential threats
- Acceptance criteria:
    - Given the identified security risks
    - When implementing the integration
    - Then all recommended security measures should be put in place

User story 3 – conduct performance testing:

- Goal: As a performance tester, I want to conduct performance testing on the integration so that it meets the required performance standards and doesn't hinder the user experience
- Acceptance criteria:
    - Given the integration is developed
    - When conducting performance tests
    - Then the integration must meet the predefined performance benchmarks |
|---|---|

User story 4 – implement performance enhancements:

- Goal: As a developer, I want to make necessary adjustments based on performance testing feedback so that the integration operates at optimal efficiency

- Acceptance criteria:

  - Given the performance test results

  - When making adjustments

  - Then all performance issues must be resolved to meet or exceed the benchmarks

User story 5 – assess compliance requirements:

- Goal: As a compliance officer, I want to evaluate the integration's alignment with legal and regulatory requirements so that the integration is compliant with all applicable laws and standards

- Acceptance criteria:

  - Given the need to assess compliance

  - When evaluating the integration

  - Then all legal and regulatory requirements pertinent to the integration must be identified and addressed

User story 6 – implement compliance measures:

- Goal: As a developer, I want to integrate the necessary compliance controls so that the integration is fully aligned with legal and regulatory requirements

- Acceptance criteria:

  - Given the compliance requirements

  - When implementing the integration

  - Then all required controls and measures should be implemented to ensure full compliance

By structuring the user stories this way, we ensure that the security, performance, and compliance aspects are not just identified but are also actioned in development. Each of these stories is precise, focused, and meets the INVEST criteria, laying a clear path for developers, testers, and compliance officers to collaborate effectively on this critical aspect of the integration.

User training and documentation

The epic for training and documentation is thin on detail. This means the prompting process has two steps. The first is to ask ChatGPT for some guidance on what documents are recommended for a Salesforce project and the second prompt will be to create user stories for the documents:

Prompt	What standard documents are produced for a Salesforce project?
Response	For a Salesforce project, there is a comprehensive set of standard documents that are often produced to ensure clear communication, effective planning, proper implementation, and alignment with business objectives. Here's an overview of those standard documents:
	1. Project charter: This outlines the purpose, scope, objectives, stakeholders, and approach of the project. It serves as a formal agreement for the project's initiation.
	2. Business requirements document (BRD): This captures the business needs, goals, and the context for the project. It may include detailed user stories or use cases.
	3. Technical design document (TDD): This is vital for developers and contains the technical specifications, architecture, data models, and any integration details. It may include details specific to Salesforce, such as Apex Classes, Triggers, Visualforce Pages, Lightning Components, and more.
	4. Functional design document (FDD): This focuses on the functional requirements, explaining how the application should work from a user's perspective.
	5. Test plan and test cases: This includes the strategy for testing along with detailed test cases, outlining what will be tested, how it will be tested, and the expected outcomes.
	6. Deployment plan: A guide for how the solution will be moved through different Salesforce environments (such as development, QA, UAT, and production), including the migration of metadata and data.
	7. User guides and training materials: These provide end users with instructions on how to use the system and can be crucial for user adoption.
	8. Security and compliance documents: These outline the security measures, data privacy considerations, and regulatory compliance related to the Salesforce solution.
	9. Performance metrics and dashboards: These detail the key performance indicators (KPIs) and may include actual Salesforce dashboards for monitoring.
	10. Data migration plan: If applicable, this document outlines how data will be migrated into or within Salesforce, including mapping, transformation, and validation.
	11. Maintenance and support guides: These cover the ongoing support and maintenance of the Salesforce solution, including any service-level agreements (SLAs).

	12. Project closure report: A final document to summarize the project, noting achievements, lessons learned, and any post-project recommendations.
	13. Governance framework: This document can be crucial for larger or complex Salesforce implementations, describing the governance structure, roles, and responsibilities for ongoing decision-making and evolution of the Salesforce environment.
	14. Disaster recovery plan: This outlines the steps and processes to follow in the event of a system failure or other catastrophic event.
	These documents encompass various stages of a Salesforce project and are tailored to the different roles involved, ensuring that everyone from stakeholders and project managers to developers and end users are aligned and informed. Depending on the size, complexity, methodology, and specific needs of the project, some documents may be combined or further subdivided.

Now that there is a list of project documents, let's get ChatCPT to write a user story for each document.

Prompt	Please write the user stories for the following epic and each of the project documents in the previous prompt/response about Salesforce project documentation. 8. User training and documentation: • Develop comprehensive training materials and documentation to ensure that sales reps can effectively utilize the new features.

Response	Certainly! Here are the user stories corresponding to each of the 13 recommended documents typically produced for a Salesforce project. Each user story is tailored to meet the INVEST criteria.
	
	User story 1 – project charter:
	
	    • Goal: As a project stakeholder, I want a project charter so that I understand the objectives, scope, stakeholders, and overall plan for the project
	
	    • Acceptance criteria:
	
	        ▪ Given that the project has been initiated
	
	        ▪ When I review the initial project documentation
	
	        ▪ Then the project charter should detail the key aspects of the project, including objectives, scope, stakeholders, and plan
	
	User story 2 – requirement documents:
	
	    • Goal: As a business analyst, I want comprehensive requirement documents so that I understand the functional and non-functional requirements of the project

- Acceptance criteria:

  • Given the project's goals and objectives

  • When I review the requirement documents

  • Then they should outline the functional and non-functional requirements clearly

User story 3 – technical design documents:

- Goal: As a developer, I want technical design documents so that I understand how the solution should be architecturally designed and implemented

- Acceptance criteria:

  • Given the project requirements

  • When I review the technical design documents

  • Then I should find a complete architectural design for the solution

User story 4 – test plan:

- Goal: As a QA engineer, I want a test plan so that I know how to structure and execute testing throughout the project

- Acceptance Criteria:

  • Given the project's design and development stages

  • When I review the test plan

  • Then I should find a comprehensive strategy for testing the solution, including test cases, criteria, and schedules

User story 5 – test cases:

- Goal: As a QA engineer, I want detailed test cases so that I can execute specific tests to validate each requirement

- Acceptance criteria:

  • Given the test plan

  • When I review the test cases

  • Then I should find specific scenarios, input data, expected results, and execution procedures for each test

User story 6 – deployment plan:

- Goal: As a DevOps engineer, I want a deployment plan so that I understand the procedures and environments for deploying the solution

- Acceptance criteria:

  - Given the project's development phase

  - When I review the deployment plan

  - Then I should find the detailed steps, environments, and procedures for deploying the solution

User story 7 – user guides:

- Goal: As a sales rep, I want detailed user guides so that I understand how to use the implemented features

- Acceptance criteria:

  - Given that the features have been implemented

  - When I access the documentation section

  - Then I should find comprehensive user guides for each feature

User story 8 – training materials:

- Goal: As a sales rep, I want hands-on training materials so that I can learn to use the new features effectively

- Acceptance criteria:

  - Given that I need to learn the new features

  - When I access the training materials

  - Then I should find various resources such as videos, tutorials, and practice exercises

User story 9 – maintenance plan:

- Goal: As a system administrator, I want a maintenance plan so that I understand how to support and maintain the solution post-deployment

- Acceptance criteria:

  - Given the deployed solution

  - When I review the maintenance plan

- Then I should find strategies and schedules for ongoing support and maintenance

User story 10 – security and compliance documentation:

- Goal: As a security analyst, I want security and compliance documentation so that I know how the project adheres to regulatory and security standards

- Acceptance criteria:

  - Given the need for compliance with regulations

  - When I review the security and compliance documentation

  - Then I should find detailed information about how the project meets relevant legal and security requirements

User story 11 – change management plan:

- Goal: As a project manager, I want a change management plan so that I understand how changes to the project scope, timeline, or resources are handled

- Acceptance criteria:

  - Given the dynamic nature of the project

  - When I review the change management plan

  - Then I should find procedures for requesting, evaluating, and implementing changes during the project

User story 12 – risk management plan:

- Goal: As a project manager, I want a risk management plan so that I know how to identify, evaluate, and mitigate risks throughout the project

- Acceptance criteria:

  - Given the project's complexities and uncertainties

  - When I review the risk management plan

  - Then I should find strategies for managing potential risks and uncertainties

User story 13 – project closure report:

- Goal: As a project stakeholder, I want a project closure report so that I can evaluate the project's performance, lessons learned, and final deliverables

- Acceptance criteria:

  - Given the project's completion

  - When I review the project closure report

  - Then I should find an assessment of the project's outcomes, performance metrics, and a summary of lessons learned

These user stories cover the spectrum of documents that are essential for a successful Salesforce project, offering clarity and direction to all team members involved.

Appendix B:
A Deep Dive Into ChatGPT and User Stories

User stories have become the lingua franca of defining development on many modern IT projects, serving as a bridge between technical complexity and human understanding. They capture the essence of a requirement in a manner that is both concise and relatable to various stakeholders, from developers to business analysts. While it's not the only way to frame prompts for generative AI tools such as ChatGPT to aid in development, the widespread familiarity and acceptance of user stories make them a logical choice for interaction. Team members and stakeholders are accustomed to thinking and communicating through this format, making it a coherent method to engage with AI as well. Recognizing this synergy, this book primarily focuses on using user stories as a means to articulate and translate requirements. This appendix provides a guide on how to utilize ChatGPT effectively to convert requirements into user stories, thereby maintaining a familiar and consistent approach to the development process.

Defining the process

One of the distinctive advantages of utilizing ChatGPT in the development process is its ability to understand and generate content based on examples rather than requiring explicit instructions. In the context of translating requirements into user stories, this ability proves invaluable. Rather than investing time and effort in explaining how to structure and write user stories from given requirements, all that is needed is to provide ChatGPT with some illustrative examples of the desired outcome. The AI can then infer the underlying pattern and create user stories that align with the provided examples. This approach not only simplifies the process but also fosters a more intuitive and adaptive workflow. It allows developers and analysts to focus more on the content and quality of the requirements, knowing that the translation into user stories can be handled efficiently by AI:

Prompt	I am going to ask you to help me turn natural language requirements into structured user stories. The first step in this process is going to be me providing you with some sample user stories. Please analyze the user stories for both content, structure, and level of detail. Do not write a response attempting to "answer" my prompt, simply ingest the materials and prepare to be asked questions.
Response	Understood. I'm prepared to analyze the user stories you provide for their content, structure, and level of detail. Please proceed with providing the sample user stories, and I will be ready to assist you with your subsequent questions and requests.

Sample user stories

It's important to recognize that ChatGPT can be tailored to craft stories in the specific style employed within your project. However, for cohesiveness and clarity in this appendix, the focus will be on generating user stories in the style that was demonstrated in the case study found in *Appendix A*. This approach ensures that the examples and guidelines provided here align with a consistent framework, enabling you to follow along and apply the same principles to your unique project context, even as the methodology presented here serves as a foundation rather than a rigid template.

Prompt	I am going to provide you with the requirements and ask you to write user stories from these requirements. The first step in this process is to give you some sample user stories. The samples are as follows. Please ask any questions you have about the user stories and share your thoughts on how the structure and approach can be improved.
	User story 1 – create a Snowflake Section within the Salesforce account pages:
	• **Goal**: As a sales rep, I want a dedicated "Snowflake" section within the Salesforce account pages so that I can quickly access the Snowflake data related to an account
	• **Acceptance criteria**:
	• **Given** a sales rep is viewing a Salesforce account page
	• **When** the page is loaded
	• **Then** a dedicated "Snowflake" section should be displayed
	L2 user story 1.1 – design the Snowflake section's layout:
	• **Goal**: As a sales rep, I want the "Snowflake" section to have a consistent and user-friendly layout, allowing for clear visibility of different data categories

	• **Acceptance criteria**:
	· **Given** a sales rep is viewing the "Snowflake" section
	· **When** the page is loaded
	· **Then** the layout of the section should be visually appealing and intuitive
	User story 2 – display billing history in the Snowflake section:
	• **Goal**: As a sales rep, I want to see the billing and payment history analysis from Snowflake within the "Snowflake" section so that I can understand the account's financial status
	• **Acceptance criteria**:
	· **Given** a sales rep is viewing the "Snowflake" section
	· **When** the billing history data is retrieved from Snowflake
	· **Then** the billing history should be displayed in a dedicated text field
	User story 3 – display product analysis in the Snowflake section:
	• **Goal**: As a sales rep, I want to see products quoted but not purchased, as analyzed by Snowflake within the "Snowflake" section, so that I can understand opportunities for up-selling or cross-selling
	• **Acceptance criteria**:
	· **Given** a sales rep is viewing the "Snowflake" section
	· **When** the product analysis data is retrieved from Snowflake
	· **Then** the analysis should be displayed in a dedicated text field
Response	Thank you for providing these sample user stories. Analyzing them, I can observe a consistent structure and approach that helps in conveying the requirements clearly. The user stories are well-structured, with a clear goal followed by acceptance criteria that detail the specific functionality desired. Here's a detailed analysis and thoughts on potential improvements:
	1. Consistency and clarity: The structure is consistent across all the stories, which is a positive aspect. Each story starts with a goal that defines the user and their need, followed by acceptance criteria that outline the expected behavior.
	2. Layering of user stories: Including an L2 user story for a subtask (design the Snowflake section's layout) adds a layer of granularity that can be beneficial for development. It might be beneficial to consistently use this approach where needed.

3. **Explicit mention of actors**: Each story mentions the user role ("sales rep") that will interact with the feature. This provides context and helps in understanding the perspective from which the requirements are being considered.

4. **Acceptance criteria format**: The Given-When-Then format that's used for acceptance criteria is concise and aligns with common practices in Agile methodologies.

5. **Potential area for improvement**: While the provided stories are clear, they might benefit from adding non-functional requirements where applicable. This can include performance, security, or usability considerations that might affect the implementation.

6. **Questions and clarifications**:

- Is there a specific style guide or template that should be followed for additional layers (L3, L4, and so on) if more granularity is required?

- Should any technical constraints or dependencies be mentioned within the user stories, or are these considered separately?

In summary, these user stories provide a clear and concise representation of the desired features, with a logical structure that could easily be followed or adapted for further requirements. However, considering non-functional aspects and clarifying any specific stylistic or content guidelines could further refine the approach.

The collaborative process of working with ChatGPT, particularly in the context of writing user stories from requirements, demands active engagement, critical thinking, and iterative refinement. The initial response from ChatGPT is often a starting point that provides a broad outline or approach, and the user's interaction with ChatGPT can significantly shape and fine-tune the outcome.

For example, when dealing with complex requirements involving **non-functional requirements** (**NFRs**) and hierarchically decomposing user stories into L1, L2, and L3 levels, a seasoned professional's insights can help guide ChatGPT to produce more aligned, nuanced, and effective user stories. By critically evaluating the feedback from ChatGPT, identifying gaps or areas for improvement, and providing refining prompts, the user not only enhances the quality of the user stories but also steers the AI toward a deeper understanding of the specific project context.

This iterative dialogue between the user and ChatGPT reinforces the notion that AI is a development partner rather than a one-size-fits-all solution. It's a dynamic relationship that thrives on continuous feedback and adjustment. ChatGPT's efficacy in such a complex task hinges on the quality of interaction, clear guidance, and a user's willingness to engage in a back-and-forth exploration of ideas. The user's expertise, ability to ask the right questions, and inclination to dig deeper play a critical role in unlocking the full potential of ChatGPT, enabling it to align with the unique needs and constraints of a given project:

Prompt	Thank you for the feedback. NFRs will be provided to you separately when appropriate. For L2 requirements, please use the convention user story 1.1, user story 1.1.1, and so forth, with each period marking the transition to the next level. Yes, please note constraints and dependencies as part of the user story.
Response	Certainly! Your guidelines provide a clear direction for structuring the user stories with additional layers and including relevant constraints and dependencies. This approach allows for a more comprehensive view of the requirements, fostering a well-rounded understanding for the development team.
	The incorporation of NFRs separately adds an additional layer of detail to the requirements process, ensuring that key considerations such as performance, security, and usability are addressed. This separation can aid in clarity, allowing for focused attention on functional and non-functional aspects individually.
	The convention of using periods to mark the transition to the next level (user story 1.1, user story 1.1.1, and so on) brings a systematic and hierarchical approach to the requirements. This will be instrumental in breaking down complex features into smaller, more manageable parts, allowing for better planning and execution.
	Noting constraints and dependencies within the user story ensures that all pertinent information is encapsulated within the story itself, allowing for better assessment and alignment with other components of the system. This complete view will facilitate more accurate planning and implementation.
	In summary, these guidelines provide a solid foundation for translating requirements into structured user stories that are coherent and comprehensive, encompassing functional and non-functional requirements, constraints, and dependencies. This well-defined approach aligns with best practices in Agile methodologies and is well-suited to driving successful project execution.

Requirements for user stories

Setting the stage when prompting ChatGPT for any project task, especially in a complex endeavor such as converting requirements into user stories, is a fundamental step that significantly impacts the outcome. Engaging with ChatGPT as a partner rather than merely a tool means taking the time to define, refine, and communicate the context, expectations, and specifics of what is needed. This interaction sets up a more coherent and aligned dialogue between the human expert and AI, allowing for a more nuanced and contextually appropriate response.

When working on converting requirements into user stories, starting with well-crafted prompts is key. Utilizing ChatGPT's expertise to help in formulating high-quality and optimized prompts sets the right tone for the entire workflow. It helps in understanding the requirements, translating them into actionable user stories, and aligning them with the overall project goals. This approach recognizes ChatGPT not merely as a source of immediate answers but as an integral part of a methodical, iterative, and collaborative development process.

This may seem somewhat meta as it involves using ChatGPT to refine the very prompts that will then be used to ask it for further assistance. However, this is precisely what allows for a deeper, more meaningful engagement. It recognizes the subtleties and complexities of the task at hand and treats ChatGPT as a dynamic contributor to the process, rather than a static responder to isolated queries. It's this kind of mindful, intentional interaction that enables the fullest and most productive utilization of AI in modern project development, ensuring alignment with both technical requirements and broader strategic objectives:

Prompt	I want to write a standard prompt prefix for asking you to ingest a requirement, ask any needed questions, and then write a great user story using the agreed-upon structure and approach. My first attempt at the prompt prefix is as follows. How can this prefix be improved so that it is very clear to you what I want done?
	Please write a user story or user stories, using the agreed-upon user story template and approach, for the following requirement. Please ask any questions needed to clarify the intent of this prompt.
Response	Your prompt prefix is clear and to the point, but adding a bit more specificity could further enhance the clarity and guidance for the task. Here's a revised version:
	"Based on the requirement provided here, please analyze and write one or more user stories following our agreed-upon structure (including hierarchical levels, constraints, and dependencies, if applicable). Use the Goal, Acceptance Criteria, and so on template, as demonstrated in previous examples. Should you need any clarification or additional information regarding the requirement's intent, context, or specific details, please ask the necessary questions before proceeding with the user story creation."
	This revision maintains the essence of your original prompt but adds some details to ensure that all aspects of the agreed-upon approach are considered. It emphasizes the need for understanding the requirement fully and encourages inquiry if there's any ambiguity, thus ensuring alignment with your expectations.

With the recommended "please write user stories from requirements" prefix firmly established, it's time to delve into the vital next stage of combining this prefix with the various specific requirements. This prefix, which has been carefully crafted and tailored to the project's needs, serves as a foundation, setting a clear direction for translating individual requirements into actionable user stories. It's an essential preamble that helps both align the AI's understanding with the task at hand and standardize the way requirements are converted across different parts of the project.

Combining this prefix with the diverse array of requirements ensures consistency and cohesion in the user stories that emerge. It sets a unifying thread that ties together different facets of the project, ensuring that each requirement is approached with the same methodology and within the same overarching context. This approach not only streamlines the process but also enhances the quality and alignment of the user stories with the project's goals. With this foundational work in place, the stage is set for a seamless, efficient, and effective translation of requirements into user stories, marking a significant step in the development process. It's a thoughtful approach that reflects a deep understanding of both the technical aspects of Salesforce development and the nuanced, collaborative nature of working with a tool such as ChatGPT:

Prompt	Based on the requirement provided here, please analyze and write one or more user stories following our agreed-upon structure (including hierarchical levels, constraints, and dependencies, if applicable). Use the Goal, Acceptance Criteria, and so on template, as demonstrated in previous examples. Should you need any clarification or additional information regarding the requirement's intent, context, or specific details, please ask the necessary questions before proceeding with the user story creation. Add a custom Boolean field called "VIP Account" to the Salesforce Account object and add it to the bottom of the Details tab on the "Sales Rep" page layout.
Response	Certainly! The requirement you provided appears to be focused on enhancing the Salesforce Account object with a custom Boolean field and modifying the "Sales Rep" page layout. Here's the user story capturing this requirement: User story – add a VIP Account field to the Salesforce Account object: • **Goal**: As a sales rep, I want a custom Boolean field named "VIP Account" added to the Salesforce Account object and displayed at the bottom of the Details tab on the "Sales Rep" page layout. This will allow me to easily identify and manage VIP accounts. • **Acceptance criteria**: ▪ **Given** a sales rep is viewing the Salesforce Account object ▪ **When** the account details are loaded ▪ **Then** a custom boolean field named "VIP Account" should be present ▪ **Given** a sales rep is viewing the "Sales Rep" page layout ▪ **When** the Details tab is selected ▪ **Then** the "VIP Account" field should be displayed at the bottom of the tab This user story should encompass the requirements you provided. Please confirm if it aligns with your expectations or if any additional details or modifications are needed.

The user story we created earlier focuses on adding the custom Boolean field "VIP Account" and how it's displayed within the Sales Rep page layout. However, it doesn't address the permissions aspect, specifically the read permissions for the Sales Rep profile on this new field. To ensure that the Sales Rep profile has the proper permissions to view and interact with this field, an additional acceptance criterion or a separate user story should be formulated

Prompt	Based on our previous exchange regarding the addition of the "VIP Account" field, I realize that we need to consider the permissions for the Sales Rep profile on this new field. Please add an acceptance criterion to the existing user story or create a new user story to address the requirement that the Sales Rep profile should be granted read permissions on the "VIP Account" field. Be sure to include the necessary details to accurately define the context and expectations.
Response	Certainly! Given the importance of ensuring that the Sales Rep profile has read permissions on the new "VIP Account" field, we can augment the existing user story with an additional acceptance criterion. This ensures that the requirement is encompassed within the context of the creation and placement of the new field.

Here's an extension to the existing user story:

User story – add a "VIP Account" field to the Salesforce Account object:

- **Goal**: As a sales rep, I want to have a "VIP Account" Boolean field on the Salesforce Account object, accessible on the Sales Rep page layout, so that I can identify VIP accounts quickly

- **Acceptance criteria**:

  - **Given** a sales rep is viewing the Details tab on the Sales Rep page layout

  - **When** the page is loaded

  - **Then** the "VIP Account" Boolean field should be displayed at the bottom of the Details tab

  - **Given** the new "VIP Account" field has been added

  - **When** the permissions are being configured

  - **Then** the sales rep profile should be granted read permissions on the "VIP Account" field

This addition ensures that not only is the field created and displayed as intended, but that the necessary permissions are also configured for the Sales Rep profile, aligning with the overall goal of the user story. It brings a clear understanding of what is expected from both a functional and a security perspective.

ChatGPT is a partner in the iterative process of crafting and refining requirements. This collaborative approach echoes the dynamic interaction often seen between development team members, with the added advantage of instantaneous access to a wealth of knowledge and insight.

Take, for example, the process of converting a requirement into a user story, as demonstrated in the creation of the "VIP Account" field user story. The initial iteration provided a solid foundation, capturing the essential elements of the requirement. However, it was through the interaction between the human expert and ChatGPT that a missing element – the permissions configuration – was identified. This realization prompted a subsequent iteration, refining the user story to include the necessary details about permissions for the Sales Rep profile.

This iterative engagement with ChatGPT is not a deviation from standard practice but rather a reflection of how user story writing is meant to function. It mirrors the back-and-forth dialogue that's typically experienced within a development team, where continuous refinement and reassessment are key. Through these iterative cycles, both clarity and completeness are achieved, ensuring that the final user story aligns with the requirement's true intent and context. This collaborative approach with ChatGPT facilitates a more comprehensive understanding and execution of the development process, thereby aligning with the principles of agile development and enhancing the efficiency and effectiveness of the development journey.

Prompt	Based on the requirement provided here, please analyze and write one or more user stories following our agreed-upon structure (including hierarchical levels, constraints, and dependencies, if applicable). Use the Goal, Acceptance Criteria, and so on template, as demonstrated in previous examples. Should you need any clarification or additional information regarding the requirement's intent, context, or specific details, please ask the necessary questions before proceeding with user story creation.
	When a Salesforce lead record is created, check if there are any other open leads with the same phone number or email address. If there are open leads with the same phone number or email address, append "Alert! There is another open lead with the same phone number or email address!" to the standard Description field of all of the identified leads.
Response	Certainly! Let's break down the requirements and craft user stories that will accurately depict the functionalities needed.
	User story 1 – identify duplicate leads by phone number or email address:
	• **Goal**: As a sales rep, I want the system to check for any other open leads with the same phone number or email address when a new lead record is created so that I can be alerted of potential duplicates
	• **Acceptance criteria**:
	▪ **Given** a new lead record is being created
	▪ **When** the lead's phone number or email address is entered

- **Then** the system should search for any open leads with the same phone number or email address

User story 1.1 – append an alert message to identified leads:

- **Goal**: As a sales rep, I want an alert message appended to the standard Description field of all identified leads with the same phone number or email address so that I can easily recognize the potential duplicate leads

- **Acceptance criteria**:

  - Given the system identifies one or more open leads with the same phone number or email address as the new lead

  - When the lead record is saved

  - Then the system should append "Alert! There is another open lead with the same phone number or email address!" to the standard Description field of all identified leads

Dependencies:

User story 1.1 is dependent on user story 1 as duplicate leads must be identified before the alert message is appended to the Description field.

Constraints:

- The alert message should only be appended to open leads, not closed ones

- The alert message should not be duplicated if the process is run on the same lead again

These user stories cover the main functionalities required to detect duplicate leads by phone number or email address and append an alert message to the identified leads. Do you have any specific constraints, dependencies, or additional information that needs to be considered?

In the context of crafting user stories, it's essential to understand that an L1 user story with L2 children may not itself be a story that gets developed. Instead, it may serve as an overarching theme or an umbrella for more specific L2 stories that break down the requirements into actionable development tasks. Only user stories that are left without children usually get developed as they represent distinct functionalities or features that can be implemented. This approach should be clearly defined in the project standards, and if not adhered to, it can lead to confusion and lack of clarity in the development process. Including guidance on this structure and explanation in the "how to write user stories" documentation is critical. ChatGPT can assist in defining and documenting these standards to ensure that everyone involved in the project shares a consistent understanding of how to structure and utilize user stories at different levels.

Prompt	Based on the requirements provided here, please analyze and write one or more user stories following our agreed-upon structure (including hierarchical levels, constraints, and dependencies, if applicable). Use the Goal, Acceptance Criteria, and so on template, as demonstrated in previous examples. Should you need any clarification or additional information regarding the requirement's intent, context, or specific details, please ask the necessary questions before proceeding with the user story creation. On each Salesforce record is a custom field called "Enterprise Account ID" that uniquely identifies the account across all IT systems used by Acme. The system of record for creating and assigning the Enterprise Account ID is Acme's Snowflake system. When a new account is created in Salesforce, the information about the account is sent to the Snowflake system to get the Enterprise Account ID to be used. When the Enterprise Account ID is received from the Snowflake system, Salesforce will check if another account is already using the same Enterprise Account ID; if it is, the new account record will be merged into the existing account record.
Response	Certainly! Let's analyze the provided requirement and transform it into well-structured user stories. The requirement indicates a multi-step process where a new Salesforce account record triggers the retrieval of an "Enterprise Account ID" from Acme's Snowflake system. Subsequently, this ID is checked against existing records in Salesforce, leading to a merge if a duplicate is found. User story 1 – retrieve the Enterprise Account ID from the Snowflake system: <ul><li>**Goal**: As a system, I want to obtain the Enterprise Account ID from the Snowflake system when a new account is created in Salesforce so that the account can be uniquely identified across all IT systems used by Acme</li><li>**Acceptance criteria**:<ul><li>**Given** a new account is created in Salesforce</li><li>**When** the creation event is triggered</li><li>**Then** a request should be sent to Acme's Snowflake system to obtain the Enterprise Account ID</li><li>**And** the received Enterprise Account ID should be stored in the custom "Enterprise Account ID" field within the Salesforce Account record</li></ul></li></ul>User story 2 – check and merge the duplicate Enterprise Account ID: <ul><li>**Goal**: As a system, I want to ensure that the Enterprise Account ID is unique within Salesforce so that if a new account record is found to have the same Enterprise Account ID as an existing record, the new record will be merged into the existing one</li></ul>

- **Acceptance criteria:**

  - **Given** the Enterprise Account ID is received from the Snowflake system for a new account

  - **When** the system checks the existing account records in Salesforce

  - **And** an account with the same Enterprise Account ID is found

  - **Then** the new account record should be merged into the existing account record with that Enterprise Account ID

  These user stories encapsulate the different stages of the requirement, breaking it down into actionable tasks. The hierarchical structure can be further expanded if more granular details are needed for development. Should you need more information or clarification, please let me know!

ChatGPT functions as a partner in the development process, extending the capability of the development team but not replacing human expertise and insight. Your understanding of the Salesforce environment and the specific integration needs with Acme's Snowflake system are critical to developing a robust and scalable solution. The given example demonstrates the complexity of real-world scenarios, where many intricacies must be considered and planned for.

In the case of the interaction between Salesforce and the Snowflake system to retrieve the Enterprise Account ID, an asynchronous design may be necessary. This would entail setting up a callout, designing a web service, implementing account record locking, and notifying the user of specific outcomes, such as the merging of an account record. These complex tasks require L2 user stories that detail how each step will be carried out.

Complex enterprise environments often come with a multitude of integrations, and it is not uncommon to employ middleware such as Mulesoft to facilitate these integrations. While such middleware may appear to simplify the integration process, in reality, it often introduces additional layers of complexity. Understanding this complexity and navigating it effectively requires a nuanced approach that considers the intricate web of interdependencies and communication channels that emerge.

To draw upon the language used by Fred Brooks in *The Mythical Man-Month*, let's consider a scenario where the Salesforce team needs to work with Snowflake. Without middleware, a direct communication channel is required between the two parties. Introducing Mulesoft to the mix, however, doubles the communication channels as the Salesforce team must now coordinate with both the Snowflake team and the Mulesoft team. This increases the complexity, not just mathematically but also operationally. The integration is no longer a straightforward connection but a multifaceted relationship involving multiple technologies and teams.

The increase in complexity translates to twice as many meetings and twice as much coordination. Managing these relationships and ensuring that all parties are aligned and working toward a common goal becomes a project in itself. Each team comes with its unique perspective, technical expertise, and constraints. Finding common ground and developing a cohesive, integrated solution demands a higher level of collaboration and communication.

This is where tools such as ChatGPT can provide tremendous value. By offering insights and background information on the various technologies involved, ChatGPT can help prepare the Salesforce team for these more complex interactions. Having an informed point of view and a solid understanding of the broader technical landscape empowers the team to engage effectively in discussions and make informed decisions.

Prompt	I am a Salesforce developer. I have a requirement to send account information to a Snowflake system and get back an asynchronous response. Our project uses Mulesoft. To prepare for a kickoff meeting with the Mulesoft team, I would like to understand how they might tackle this requirement. The Mulesoft team is using the Mulesoft Salesforce Connector to connect to Salesforce. Please write the user stories from the point of view of the Mulesoft team.
	Based on the requirement provided here, please analyze and write one or more user stories following our agreed-upon structure (including hierarchical levels, constraints, and dependencies, if applicable). Use the Goal, Acceptance Criteria, and so on template, as demonstrated in previous examples. Should you need any clarification or additional information regarding the requirement's intent, context, or specific details, please ask the necessary questions before proceeding with user story creation.
	Receive information about a new account in Salesforce and send the information to the Snowflake system to get an "Enterprise Account ID." When the Enterprise Account ID is received from the Snowflake system, update the custom Enterprise Account ID on the Salesforce account record.
Response	Certainly! Since the perspective is that of the Mulesoft team, the user stories will focus on the integration process, specifically handling the reception of information from Salesforce, the communication with the Snowflake system, and the update back to Salesforce. The Mulesoft Salesforce Connector plays a crucial role in this scenario. User story 1 – establish a connection with Salesforce: • **Goal**: As a Mulesoft developer, I want to establish a seamless connection with Salesforce using the Mulesoft Salesforce Connector to ensure that the new account information is received correctly

- **Acceptance criteria**:

  - **Given** the Mulesoft Salesforce Connector is configured

  - **When** a new account is created in Salesforce

  - **Then** the connector must successfully receive the account information and log the event

L2 user story 1.1 – validate Salesforce account information:

- **Goal**: As a Mulesoft developer, I want to validate the received account information to ensure that it's correctly formatted before sending it to the Snowflake system

- **Acceptance criteria**:

  - **Given** the account information is received from Salesforce

  - **When** it's parsed and validated

  - **Then** the system must confirm its readiness to send to the Snowflake system

User story 2 – send account information to the Snowflake system:

- **Goal**: As a Mulesoft developer, I want to transmit the validated account information to the Snowflake system so that the Enterprise Account ID can be generated

- **Acceptance criteria**:

  - **Given** the validated account information

  - **When** it's sent to the Snowflake system

  - **Then** a successful transmission confirmation must be logged, and the system must be prepared to receive an asynchronous response

User story 3 – receive the Enterprise Account ID from Snowflake:

- **Goal**: As a Mulesoft developer, I want to correctly receive the asynchronous response from the Snowflake system containing the Enterprise Account ID

- **Acceptance criteria**:

  - **Given** the Snowflake system sends the Enterprise Account ID

  - **When** it's received by Mulesoft

  - **Then** it must be validated and prepared for updating in Salesforce

L2 user story 3.1 – update the Salesforce account record:

- **Goal**: As a Mulesoft developer, I want to update the custom Enterprise Account ID on the Salesforce account record using the Mulesoft Salesforce Connector

- **Acceptance criteria**:

  - **Given** the Enterprise Account ID is received and validated

  - **When** it's sent to Salesforce through the connector

  - **Then** the Salesforce account record must be successfully updated, and a confirmation of the update must be logged

These user stories provide the Mulesoft team with a clear path to handle the integration requirement from their point of view, focusing on the essential interactions between Salesforce, Mulesoft, and Snowflake. If further details or specific aspects of this process need to be explored, please let me know, and we can dive deeper into the complexities of this integration scenario.

This response emphasizes an approach where the Mulesoft team takes the reins in managing the integration between Mulesoft and Salesforce. This is a strategic decision that leverages specific tools and expertise, particularly the Mulesoft Salesforce Connector, to streamline the process.

Utilization of the Mulesoft Salesforce Connector is a key aspect of this approach. The Mulesoft Salesforce Connector is a specialized tool that's designed to facilitate the integration between Mulesoft and Salesforce. By allowing the Mulesoft team to take control of this integration, they can fully leverage this tool's capabilities. The Mulesoft team's familiarity with the connector means they can navigate its features efficiently, optimizing the integration process.

Another important element of this approach is minimizing the Salesforce team's involvement. Since the Mulesoft team will handle the bulk of the hands-on integration work, the Salesforce team's involvement is significantly reduced. This can be beneficial, especially if the Salesforce team lacks specific expertise in Mulesoft or if their resources are better allocated elsewhere. The Mulesoft team's specialized knowledge enables them to address the integration's nuances and complexities without needing to rely heavily on the Salesforce team's input.

The proposed approach also aligns with a best practice in complex integrations: leveraging specialized expertise where it exists. By entrusting the integration to the team with the most experience in the particular technologies involved, the process is likely to be more efficient and less error-prone. This specialized focus on integration often translates into faster development times and a more robust end product.

While the approach has clear benefits, it's also essential to recognize potential challenges. Communication between the Mulesoft and Salesforce teams must be clear and ongoing to ensure alignment with business requirements and overall project goals. The reduced involvement of the Salesforce team doesn't negate the need for collaboration; rather, it underscores the importance of clear roles, responsibilities, and communication channels.

To select this approach, it must align with the broader business and project goals. It is not a one-size-fits-all solution but rather a strategic decision based on the specific context, needs, and expertise of the teams involved. Understanding how the approach fits into the larger picture is crucial for its successful implementation.

Prompt	Assuming that the Mulesoft Salesforce Connector is already deployed and in use, these Mulesoft stories would seem to require little to no work from Salesforce. Is this a correct assumption?
Response	Yes, your assumption is mostly correct. Given that the Mulesoft Salesforce Connector is already deployed and in use, the integration process should primarily be handled by the Mulesoft team. They will be responsible for receiving the account information from Salesforce, interacting with the Snowflake system, and then updating the Salesforce account record with the Enterprise Account ID.
	However, it's important to consider a few areas where collaboration or coordination with the Salesforce team might be required:
	1. Data format and schema: Ensure that the data format and schema align with what the Mulesoft process expects. If specific fields or a particular structure are needed, collaboration with the Salesforce team might be essential to ensure consistent data quality.
	2. Triggering mechanism: Depending on how the integration is triggered (for example, a real-time event when an account is created), there might need to be some alignment with Salesforce to ensure the triggering mechanism is properly configured.
	3. Security and Permissions: Ensure that the Mulesoft Salesforce Connector has the appropriate permissions to both read the necessary account information and update the custom Enterprise Account ID field in Salesforce. This may involve some configuration on Salesforce's side.
	4. Testing and validation: During the development and testing phases, collaboration between the Mulesoft and Salesforce teams might be required to validate that the integration is working as intended, especially if any changes or updates are made to related objects or processes in Salesforce.
	While the bulk of the development work may indeed reside with the Mulesoft team, maintaining clear communication and collaboration between the teams can help ensure a smooth and successful integration. If there are any concerns about how the existing deployment of the Mulesoft Salesforce Connector might influence this requirement, additional analysis and alignment with both teams may be warranted.

ChatGPT serves as a development partner, aiding in the ideation, planning, and execution of various technical tasks. However, it's essential to recognize that it doesn't replace the nuanced understanding, creativity, and expertise that human developers bring to a project, particularly when dealing with complex or unfamiliar technologies such as Mulesoft.

When crafting user stories or working on integration requirements involving technologies that a user might not be intimately familiar with, ChatGPT can provide insights based on available information, best practices, and standard conventions. It can act as a guide, offering suggestions and helping shape the approach. However, the user must still apply critical thinking and validate these recommendations with real-world experience, project context, and perhaps consultation with experts in that specific technology. While ChatGPT's output is increasingly reliable, it may lack the latest or project-specific information that could influence the implementation.

In the rapidly evolving field of technology, where project requirements often intersect with unique configurations and specific business needs, human judgment remains invaluable. The collaboration between AI and human intelligence becomes a synergy, where ChatGPT offers a foundation and accelerates certain aspects of the process, while developers apply their understanding, intuition, and problem-solving skills to ensure alignment with project goals. This balanced partnership capitalizes on the strengths of both AI and human capabilities, creating a robust development environment, but it requires mindfulness about the limitations and a commitment to thorough review and validation.

Index

Packtpub.com

Subscribe to our online digital library for full access to over 7,000 books and videos, as well as industry leading tools to help you plan your personal development and advance your career. For more information, please visit our website.

Why subscribe?

- Spend less time learning and more time coding with practical eBooks and Videos from over 4,000 industry professionals

- Improve your learning with Skill Plans built especially for you

- Get a free eBook or video every month

- Fully searchable for easy access to vital information

- Copy and paste, print, and bookmark content

Did you know that Packt offers eBook versions of every book published, with PDF and ePub files available? You can upgrade to the eBook version at Packtpub.com and as a print book customer, you are entitled to a discount on the eBook copy. Get in touch with us at customercare@packtpub.com for more details.

At www.packtpub.com, you can also read a collection of free technical articles, sign up for a range of free newsletters, and receive exclusive discounts and offers on Packt books and eBooks.

Other Books You May Enjoy

If you enjoyed this book, you may be interested in these other books by Packt:

Salesforce End-to-End Implementation Handbook

Kristian Margaryan Jørgensen

ISBN: 978-1-80461-322-1

- Discover the critical activities in Salesforce implementation
- Address common issues faced in implementing Salesforce
- Explore appropriate delivery methodology
- Understand the importance of a change management strategy
- Govern Salesforce implementation through all its phases
- Gain insights on key activities in the continuous improvement phase
- Leverage customer 360 for analytics, AI and automation

Business Process Automation with Salesforce Flows

Srini Munagavalasa

ISBN: 978-1-83508-925-5

- Gain insights into gathering business requirements and identifying automation needs
- Identify opportunities for improving business process flows
- Translate critical steps in the business process flow and automate them using Flow Builder
- Identify different types of Salesforce Flows tailored to various scenarios
- Optimize and troubleshoot Salesforce Flows for increased efficiency
- Discover ways to implement complex business process automation through flow orchestration

Packt is searching for authors like you

If you're interested in becoming an author for Packt, please visit authors.packtpub.com and apply today. We have worked with thousands of developers and tech professionals, just like you, to help them share their insight with the global tech community. You can make a general application, apply for a specific hot topic that we are recruiting an author for, or submit your own idea.

Share Your Thoughts

Now you've finished *ChatGPT for Accelerating Salesforce Development*, we'd love to hear your thoughts! Scan the QR code below to go straight to the Amazon review page for this book and share your feedback or leave a review on the site that you purchased it from.

https://packt.link/r/1835084079

Your review is important to us and the tech community and will help us make sure we're delivering excellent quality content.

Download a free PDF copy of this book

Thanks for purchasing this book!

Do you like to read on the go but are unable to carry your print books everywhere?

Is your eBook purchase not compatible with the device of your choice?

Don't worry, now with every Packt book you get a DRM-free PDF version of that book at no cost.

Read anywhere, any place, on any device. Search, copy, and paste code from your favorite technical books directly into your application.

The perks don't stop there, you can get exclusive access to discounts, newsletters, and great free content in your inbox daily

Follow these simple steps to get the benefits:

1. Scan the QR code or visit the link below

https://packt.link/free-ebook/9781835084076

2. Submit your proof of purchase
3. That's it! We'll send your free PDF and other benefits to your email directly

i

www.ingramcontent.com/pod-product-compliance
Lightning Source LLC
Chambersburg PA
CBHW080616060326
40690CB00021B/4718